APPLIED MATHEMATICAL AND PHYSICAL FORMULAS

SECOND EDITION

Vukota Boljanovic, Ph.D.

APPLIED MATHEMATICAL AND PHYSICAL FORMULAS

SECOND EDITION

A QUICK REFERENCE FOR ENGINEERS,
TECHNICIANS, TOOLMAKERS, MECHANISTS,
STUDENTS, AND TEACHERS

Industrial Press, Inc.

Industrial Press, Inc.

32 Haviland Street, Suite 3
South Norwalk, Connecticut 06854
Tel: 203-956-5593, Toll-Free: 888-528-7852
E-mail: info@industrialpress.com

Applied Mathematical and Physical Formulas,
Pocket Reference, 2nd Edition, by Vukota Boljanovic

ISBN (print): 978-0-8311-3592-8
ISBN ePDF: 978-0-8311-9361-4
ISBN ePUB: 978-0-8311-9362-1
ISBN eMOBI: 978-0-8311-9363-8

Editorial Director: Taisuke Soda
Managing Editor: Laura Brengelman
Cover Designer: Janet Romano-Murray

10 9 8 7 6 5 4 3 2

industrialpress.com

ebooks.industrialpress.com

DEDICATION

To my grandson *Alexandar Boljanovic.*

TABLE OF CONTENTS

Table of Contents

Table of Contents

Table of Contents

Table of Contents

Table of Contents

Table of Contents

xviii

Table of Contents

PART III **PHYSICS**   **299**

Table of Contents

Table of Contents

Table of Contents

PREFACE

This comprehensive pocket reference guide gives students, engineers, toolmakers, metalworkers, and other specialists a wide range of mathematical and physical formulas in a handy format.

Compactly arranged in an attractive, unique style, this reference book has just about every equation, definition, diagram, and formula that a user might want in doing undergraduate-level physics and mathematics. Great care has been taken to present all formulas concisely, simply, and clearly. All of the information included is practical—rarely used formulas are excluded.

Thoroughly practical and authoritative, this book brings together in three parts thousands of formulas and figures to simplify review or to refresh your memory of what you studied in school. If you are in school now, and you don't have a lot of time but want to excel in class, this book will help you brush up before tests, find answers fast, learn key formulas and geometric figures, study quickly, and learn more effectively. In fact, each year, this indispensable study guides helps hundreds of thousands of students improve their test scores and final grades.

The first part of the book covers the International System of Units (the SI base units, the SI derived units,the SI prefixes, and units outside the SI that are accepted for use with the SI); metric units of measurement; U.S. units of measurements; and tables of equivalent metric and United States Customary System (USCS) units.

Preface

The second part of the book covers formulas, rules, and figures related to Algebra, Geometry, Trigonometry, Analytical Geometry, Mathematics of Finance, Calculus, and Statistics.

The third part of the book covers formulas, definitions, and figures related to Mechanics, Fluid Mechanics, Temperature and Heat, Electricity and Magnetism, Light, and Waves and Sound.

New in the Second Edition

This edition features essentially the same format and organization as the first edition. In addition to improved figures and new text, enhancements and extensions of several sections in the first edition have been included, and two new sections in Part II have been added.

- In Part I (concerning Units), a section on units of measure in precious metals has been added.

- In Part II (concerning Mathematics), the section on calculus has been expanded, and two new sections on Arithmetic and Mathematical Fundamentals for Computer Science have been added.

Students and professionals alike will find this book a very effective learning tool and reference.

Acknowledgments

I would like to thank my colleagues and users of the book for their helpful suggestions regarding improvements to the second edition, and my family and personal friends who indirectly contributed to this second edition.

—Vukota Boljanovic

GREEK LETTERS

α	A	alpha
β	B	beta
γ	Γ	gamma
δ	Δ	delta
ε	E	epsilon
ζ	Z	zeta
η	H	eta
θ	Θ	theta
ι	I	iota
κ	K	kappa
λ	Λ	lambda
μ	M	mu
ν	N	nu
ξ	Ξ	xi
o	O	omicron
π	Π	pi
ρ	P	rho
σ	Σ	sigma
τ	T	tau
υ	Υ	upsilon
ϕ	Φ	phi
χ	X	chi
ψ	Ψ	psi
ω	Ω	omega

NUMBERS π AND e

$\pi = 3.141\,592\,653\,589\,793$

$e = 2.718\,281\,828\,459\,045$

$\log_{10} e = 0.434\,294\,481\,903$

$\log_e 10 = 2.302\,585\,092\,994$

PART **I**

UNITS

Units are labels that are used to distinguish one type of measurable quantity from other types. Length, mass, and time are distinctly different physical quantities, and therefore have different unit names, such as meters, kilograms and seconds. We use several systems of units, including the metric (SI), the English (or U.S. customary units), and a number of others, which are of mainly historical interest.

This part of the book contains the following:

- International System of Units
- Metric Units of Measurement
- U.S. Units of Measurement
- Units of Measure in Precious Metals
- Tables of Equivalents

UNITS
International System of Units

INTERNATIONAL SYSTEM OF UNITS
The International System of Units, abbreviated as SI, is the modernized version of the metric system established by international agreement.

1. SI Base Units

Quantity	Name	Symbol
length	meter	m
mass	kilogram	kg
time	second	s
electric current	ampere	A
thermodynamic temperature	kelvin	K
amount of a substance	mole	mol

2. SI Derived Units

Quantity	Name	Symbol
area	square meter	m^2
volume	cubic meter	m^3
speed, velocity	meter per second	m/s
acceleration	meter per second squared	m/s^2
wave number	reciprocal meter	m^{-1}
mass density	kilogram per cubic meter	kg/m^3

UNITS
International System of Units

Continued from # 2

specific volume	cubic meter per kilogram	m^3/kg
current density	ampere per square meter	A/m^2
magnetic field strength	ampere per meter	A/m
amount-of-substance concentration	mol per cubic meter	mol/m^3
luminance	candela per square meter	cd/m^2
mass fraction	kilogram per kilogram	kg/kg

3. SI Derived Units with Special Names and Symbols

Quantity	Name	Symbol
plane angle	radian	rad
solid angle	steradian	sr
frequency	hertz	Hz
force	newton	N
pressure, stress	pascal	Pa
energy, work, quantity of heat	joule	J
power, radiant flux	watt	W

UNITS
International System of Units

Continued from # 3

electric charge, quantity of electricity	coulomb	C
electric potential difference	volt	V
capacitance	farad	F
electric resistance	ohm	Ω
electric conductance	siemens	S
magnetic flux	weber	Wb
magnetic flux density	tesla	T
inductance	henry	H
Celsius temperature	degree Celsius	$^{\circ}C$
luminous flux	lumen	lm
illuminance	lux	lx
activity of a radionuclide	becquerel	Bq
absorbed dose, specific energy, kerma	gray	Gy
dose equivalent	sievert	Sv
catalytic activity	katal	kat

UNITS
International System of Units

4. SI Derived Units Whose Names and Symbols Include SI Derived Units with Special Names and Symbols

Quantity	Name	Symbol
dynamic viscosity	pascal second	$Pa \cdot s$
moment of force	newton meter	$N \cdot m$
angular velocity	radian per second	rad/s
angular acceleration	radian per second squared	rad/s^2
heat flux density, irradiance	watt per square meter	W/m^2
heat capacity, entropy	joule per kelvin	J/K
specific heat capacity, specific entropy	joule per kilogram kelvin	$J/(kg \cdot K)$
specific energy	joule per kilogram	J/kg
energy density	joule per cubic meter	J/m^3
thermal conductivity	watt per meter kelvin	$W/(m \cdot K)$
electric field strength	volt per meter	V/m
electric charge density	coulomb per cubic meter	C/m^3

UNITS
International System of Units

Continued from # 4

electric flux density	coulomb per square meter	C/m^2
permittivity	farad per meter	F/m
permeability	henry per meter	H/m
molar energy	joule per mole	J/mol
molar entropy, molar heat capacity	joule per mole kelvin	$J/(mol \cdot K)$
exposure (X and γ rays)	coulomb per kilogram	C/kg
absorbed dose rate	gray per second	Gy/s
radiant intensity	watt per steradian	W/sr
radiance	watt per square meter steradian	$W/(m^2 \cdot sr)$

5. Units Outside the SI that Are Accepted for Use with the SI

Name	Symbol	Value in SI units
minute	min	1 min = 60 s
hour	h	1 h = 60 min = 3600 s
day	d	1 d = 24 h = 86400 s
liter	L	$1\ L = 1\ dm^3 = 10^{-3}\ m^3$
metric tone	t	$1\ t = 10^3\ kg$
bel	B	1 B = 10 dB

UNITS
International System of Units

Continued from # 5

degree (angle)	°	$1° = (\pi/180)$ rad
minute (angle)	′	$1' = (1/60)° =$ $= (\pi/10800)$ rad
second (angle)	″	$1'' = (1/60)' =$ $= (\pi/648000)$ rad
electronvolt	eV	$1\,eV = 1.60218 \times 10^{-19}\,J$
unified atomic mass unit	u	$1\,u = 1.66054 \times 10^{-27}\,kg$
astronomical unit	ua	$1\,ua = 1.49598 \times 10^{11}\,m$
nautical mile		1 nautical mile = 1852 m
knot		1 knot = 1852/3600 m/s
are	a	$1\,a = 100\,m^2$
hectare	ha	$1\,ha = 100\,a = 10^4\,m^2$
bar	bar	$1\,bar = 10^2\,kPa = 10^5\,Pa$
angstrom	Å	$1Å = 0.1\ nm = 10^{-10}\,m$
curie	Ci	$1\,Ci = 3.7 \times 10^{10}\,Bq$
rad	rad	$1\,rad = 10^{-2}\,Gy$
rem	rem	$1\,rem = 10^{-2}\,Sv$

UNITS
Metric Units of Measurement

6. SI Prefixes

Factor	Name	Symb.	Factor	Name	Symb.
10^1	deca	da	10^{-1}	deci	d
10^2	hecto	h	10^{-2}	centi	c
10^3	kilo	k	10^{-3}	milli	m
10^6	mega	M	10^{-6}	micro	μ
10^9	giga	G	10^{-9}	nano	n
10^{12}	tera	T	10^{-12}	pico	p
10^{15}	peta	P	10^{-15}	femto	f
10^{18}	exa	E	10^{-18}	atto	a

METRIC UNITS OF MEASUREMENT

The metric system was first proposed in 1791. The French Revolutionary Assembly adopted it in 1795, and the first metric standards (a standard meter bar and kilogram bar) were adopted in 1799.

7. Units of Length

Name	Symbol	Value
millimeter	mm	1 mm = 0.001 m
centimeter	cm	1 cm = 10 mm

UNITS
Metric Units of Measurement

Continued from # 7

decimeter	dm	1 dm = 10 cm
meter	m	1 m = 10 dm = 1000 mm
dekameter	dam	1 dam = 10 m
hectometer	hm	1 hm = 10 dam
kilometer	km	1 km = 10 hm =1000 m

8. Units of Area

Name	Symbol	Value
sq. millimeter	mm^2	$1\,mm^2 = 0.000001\,m^2$
sq. centimeter	cm^2	$1\,cm^2 = 100\ mm^2$
sq. decimeter	dm^2	$1\,dm^2\ = 100\,cm^2$
sq. meter	m^2	$1\ m^2 = 100\,dm^2$
sq. decameter	dam^2	$1\,dam^2 = 100\ m^2$
sq. hectometer	hm^2	$1\,hm^2 = 100\,dam^2$
sq. kilometer	km^2	$1\,km^2\ = 100\,hm^2$

9. Units of Liquid Value

Name	Symbol	Value
milliliter	mL	1 mL = 0.001 L
centiliter	cL	1 cL = 10 mL
deciliter	dL	1 dL = 10 cL
liter	L	1 L = 10 dL = 1000 mL

UNITS
Metric Units of Measurement

Continued from # 9

dekaliter	daL	1 daL = 10 L
hectoliter	hL	1 hL = 10 daL
kiloliter	kL	1 kL = 10 hL = 1000 L

10. Units of Volume

Name	Symbol	Value
cu. millimeter	mm^3	$1\ mm^3 = 10^{-9}\ m^3$
cu. centimeter	cm^3	$1\ cm^3 = 1000\ mm^3$
cu. decimeter	dm^3	$1\ dm^3 = 1000\ cm^3$
cu. meter	m^3	$1\ m^3 = 1000\ dm^3$

11. Units of Mass

Name	Symbol	Value
milligram	mg	1 mg = 0.001g
centigram	cg	1 cg = 10 mg
decigram	dg	1 dg = 10 cg
gram	g	1 g = 10 dg
dekagram	dag	1 dag = 10 g
hectogram	hg	1 hg = 10 dag
kilogram	kg	1 kg = 10 hg = 1000 g
megagram	Mg	1 Mg = 1000 kg = 1 t

U.S. UNITS OF MEASUREMENT

Most of the U.S. system of measurements is the same as that for the U.K. The biggest differences to be noted are that the present British gallon and bushel—known, as the "Imperial gallon" and the "Imperial bushel"—are, respectively, about 20 percent and 3 percent larger than the U.S. gallon and bushel.

12. Units of Length

Name	Symbol	Value
inch	in	1 in = 0.83333 ft
foot	ft	1 ft = 12 in
yard	yd	1 yd = 3 ft
rod	rd	1 rd = 16.5 ft
furlong	fur	1 fur = 40 rd
U.S. mile	mi	1 mi = 8 fur = 5280 ft
nautical mile	nautical mile	1 nautical mile = 1852 m = 6076.1149 ft (approx.)

13. Units of Area

Name	Symbol	Value
sq. inch	in^2	$1\,in^2 = 0.006444\ ft^2$
sq. foot	ft^2	$1\,ft^2 = 144\ in^2$
sq. yard	yd^2	$1\,yd^2 = 9\ ft^2$

UNITS
U.S. Units of Measurement

Continued from # 13

sq. rod	rd^2	$1 \text{ rd}^2 = 272.25 \text{ ft}^2$
acre	acre	1 acre = 160 rd $= 43\ 560 \text{ ft}^2$
sq. mile	mi^2	$1 \text{ mi}^2 = 640$ acre
township		1 township = 6 mi^2

14. Units of Liquid Volume

Name	Symbol	Value
gill	gi	1 gi = 0.25 pt
pint	pt	1 pt = 4 gi
quart	qt	1 qt = 2 pt
gallon	gal	1 gal = 4 qt = 8 pt = 32 gi

15. Units of Volume

Name	Symbol	Value
cu. inch	in^3	$1 \text{ in}^3 = 0.0005787 \text{ ft}^3$
cu. foot	ft^3	$1 \text{ ft}^3 = 1728 \text{ in}^3$
cu. yard	yd^3	$1 \text{ yd}^3 = 27 \text{ ft}^3$

UNITS
U.S. Units of Measurement

16. Apothecaries' Units of Liquid Volume

Name	Symbol	Value
minim	min	1 min = 0.016666 dr
fluid dram	fl dr	1 fl dr = 60 min
fluid ounce	fl oz	1 fl oz = 8 fl dr
pint	pt	1 pt = 16 fl oz
quart	qt	1 qt = 2 pt
gallon	gal	1 gal = 4 qt

17. Units of Dry Volume

Name	Symbol	Value
pint	pt	1 pt = 0.5 qt
quart	qt	1 qt = 2 pt
peck	pk	1 pk = 8 qt
bushel	bu	1 bu = 4 pk

18. Avoirdupois Units of Mass

Name	Symbol	Value
grain	gr	1 gr = 64.79891 mg
dram	dr	1 dr = 27-11/32 gr
ounce	oz	1 oz = 16 dr
pound	lb	1 lb = 16 oz

UNITS
U.S. Units of Measurement

Continued from # 18

hundredweight	cwt	1 cwt = 100 lb
ton	ton	1 ton = 20 cwt = 2000 lb

19. Apothecaries' Units of Mass

Name	Symbol	Value
grain	gr	1 gr = 64.79891 mg
scruple	s ap	1 s ap = 20 dr
apothecaries' dram	dr ap	1 dr ap = 3 s ap
apothecaries' ounce	oz ap	1 oz ap = 8 dr ap
apothecaries' pound	lb ap	1 lb ap = 12 lb

ap

20. Troy Units of Mass

Name	Symbol	Value
grain	gr	1 gr = 64.79891 mg
pennyweight	dwt	1 dwt = 24 gr
ounce troy	oz t	1 oz t = 20 dwt
pound troy	lb t	1 lb t = 12 oz t
pennyweight	dwt	1 dwt = 24 gr

---

21. Units of Measure in Precious Metals

The *troy ounce*, also called the fine ounce (oz t), is a measure expressing the mass of gold and other precious metals (silver and platinum).

1 fine ounce = 31.1034768 grams
32.15 fine ounces = 1 kg

The *carat* is a measure that defines the purity of gold.
Carat purity is measured as 24 times the pure mass divided by the total mass and can be calculated by the following formula:

$$K = 24 \frac{M_g}{M_m}$$

where

K = the carat rating of the precious metal
M_g = the mass of the pure precious metal
M_m = the total mass of the material.

These specific measurement units are rarely or never used outside of specialized professions such as jewelry-making and metal-investing.

UNITS
Units of Measure in Precious Metals

Comparative Table Measures the Purity of Gold

Cleanliness in carats	Cleanliness in thousands	Purity in %
24	999	99.9
22	916	91.6
21	875	87.5
20	833	83.3
18	750	75.0
15	625	62.5
14	585	58.5
10	417	41.7
1	42	4.2

Example:

If a product is made of 18 carat gold, it contains 75 percent pure gold and 25 percent of another metal, usually silver or copper.

TABLES OF EQUIVALENTS
In tables below, all bold equivalents are exact.

22. Units of Length

Name	Equivalents
1 angstrom (Å) =	**0.1** nm **0.0000001** mm 0.000000004 inch
1 centimeter (cm) =	0.3937 in
1 chain (ch) =	66 ft
1 decimeter (dm) =	3.937 in
1 dekameter (dam) =	32.808 ft
1 fathom =	**6** ft **1.8288** m
1 furlong (fur) =	**10** ch **660** ft 201.168 m
1 light-year (ly) =	9460730472580.7 km 5878625373183.61 mi
1 foot (ft) =	**0.3048** m **12** in **30.48** cm

UNITS
Tables of Equivalents

Continued from # 22

1 inch (in) =	**25.4** mm **2.54** cm
1 kilometer (km) =	0.621 mi
1 meter (m) =	39.37 in 1.094 yd
1 micrometer (μm) =	**0.001** mm
1 mile (mi) =	**5,280** ft 1.609 km
1 mile (international nautical) =	**1.852** km 1.151 mi
1 millimeter (mm) =	0.03937 in
1 nanometer (nm) =	**0.001** μm 0.000000039 37 in
1 point (typography) =	**0.013837** in 1/72 in 0.351 mm
1 rod (rd) =	**16.5** ft 5.0292 m
1 yard (yd) =	**0.9144** m

UNITS
Tables of Equivalents

23. Units of Area

Name	Equivalents
1 acre =	**43,560** ft^2 4,046 m^2 0.40467 ha
1 are (a) =	119.599 yd^2 0.025 acre
1 hectare (ha) =	2.471 acre
1 square centimeter (cm^2) =	0.155 in^2
1 square foot (ft^2) =	0.0929030 m^2
1 square inch (in^2) =	**645.16** mm^2
1 square kilometer (km^2) =	247.104 acre 0.386 mi^2
1 square meter (m^2) =	1.196 yd^2 10.764 ft^2
1 square mile (mi^2) =	258.999 ha
1 square millimeter (mm^2) =	0.002 in^2
1 square rod (rd^2) =	25.293 m^2
1 square yard (yd^2) =	0.836 m^2

UNITS
Tables of Equivalents

24. Units of Volume

Name	Equivalents
1 barrel (bbl), liquid* =	31 to 42 gal
1 bushel (bu) (U.S.) =	**2150.42** in^3 35.239 L
1 cubic centimeter (cm^3) =	0.061 in^3
1 cubic foot (ft^3) =	7.481 gal 28.316 dm^3
1 cubic inch (in^3) =	0.554 fl oz 16.387 cm^3
1 cubic meter (m^3) =	1.308 yd^3
1 cubic yard (yd^3) =	0.765 m^3
1 cup, measuring =	**8** fl oz 237 mL 0.5 pt
1 dekaliter (daL) =	2.642 gal 1.135 pk
1 hectoliter (hL) =	26.418 gal 2.838 bu
1 liter (L) =	1.057 fl qt 61.025 in^3
1 milliliter (mL) =	0.271 fl dr 0.061 in^3

UNITS
Tables of Equivalents

Continued from # 24

1 ounce, fluid (fl oz) =	1.805 in^3 29.573 mL
1 peck (pk) =	8.810 L
1 pint (pt), dry =	33.600 in^3 0.551 L
1 pint (pt), liquid =	**28.875** in^3 0.473 L
1 quart (qt), dry (U.S) =	67.201 in^3 1.101 L
1 quart (qt), liquid (U.S.) =	**57.75** in^3 0.946 L
1 dram, fluid (fl dr) =	0.226 in^3 **1/8** fl oz 3.697 mL
1 gallon (gal) (U.S.) =	**231** in^3 **128** fl oz; 3.785 L

* There are a variety of "barrels" established by
   law or usage.

25. Units of Mass

Name	Equivalents
1 carat (c) =	200 mg 3.086 gr

UNITS
Tables of Equivalents

Continued from # 25

1 dram apothecaries (dr ap) =	**60** gr 3.888 g
1 gamma (γ) =	**1** µg
1 grain (gr) =	**64.79891** mg
1 gram (g) =	15.432 gr
1 kilogram (kg) =	2.205 lb
1 ounce, troy (oz t) =	**480** gr 31.103 g
1 pennyweight (dwt) =	1.555 g
1 point =	0.01 carat 0.02 mg **7,000** gr
1 pound, troy (lb t) =	**5,760** gr 373.242 g
1 ton, net =	**2,000** lb 0.893 gross ton
1 ton, gross =	**2,240** lb **1.12** net tons 1.016 t
1 ton, metric (t) =	2,204.623 lb 0.984 gross ton 1.102 net tons

PART **II**

MATHEMATICS

Mathematics is a branch of science large enough to be distinctly separate from "science" and to be placed in its own category.

This part of the book contains the most frequently used formulas, definitions, and rules relating to the following:

- Arithmetic
- Algebra
- Geometry
- Trigonometry
- Analytical Geometry
- Mathematics of Finance
- Calculus
- Statistics
- Mathematical Fundamentals of Computer Science

ARITHMETIC

Arithmetic is one of the main branches of mathematics that deals with numbers, and it is so important that some consider it the "queen" of mathematics. The name comes from the Greek, ἀριθμός, "number."

Important early developments in arithmetic include the discovery of irrational numbers in ancient Greece, the introduction of zero and the use of positional notation in India, and the invention of the decimal system, also in India, as well as the use of Arabic numerals, developed by Hindu mathematicians.

This section contains the following:

- Fundamentals of Arithmetic
- Fractions
- Proportionality and Percentages
- Calculation for Mixtures

ARITHMETIC
Fundamentals of Arithmetic

1. Natural Numbers

Natural numbers are what we use when we are counting objects one by one. When we start using 1,2,3,4 and so on, we are using the counting numbers or, to give them their proper title, the natural numbers. The set of natural numbers, denoted by N, can be defined as:

$$N_1 = \{1, 2, 3, 4,...,n, n+1,...\} \Rightarrow \text{positive integers,}$$
$$N_0 = \{0, 1, 2, 3, 4,...,n, n+1,...\} \Rightarrow \text{non-negative integers.}$$

The set N, whether or not it includes zero, is a *denumerable* set. "Denumerability" refers to the fact that even though there might be an infinite number of elements in a set, they can be counted one by one.

a) Comparison of natural numbers
Comparing numbers in mathematics is the process of determining whether one number is equal to, smaller than, or bigger than another number. The symbols used in comparison are these:

- Greater than (>)
- Smaller than (<)
- Equal to (=)

ARITHMETIC
Fundamentals of Arithmetic

The other signs are combinations of the signs mentioned above:

- Greater than or equal to $(\geq)$
- Smaller than or equal to $(\leq)$
- Different than $(\neq)$

The easiest way to remember these signs is to remember that the point of the arrow is always on the side of the smaller number.

Example:

$$5 > 3$$
$$2 < 6$$

b) Addition of natural numbers
Addition is the mathematical operation that is signified by the plus sign (+)

$$a + b = c$$

where
$\quad a$ and b = the numbers being added, or "addends"
$\quad\quad c$ = sum

The properties of the addition of natural numbers include the following:

ARITHMETIC
Fundamentals of Arithmetic

1. *Closure*. The sum of two natural numbers is also a natural number.

$$a + b \in N$$

2. *Associativity*. The way in which the addends are grouped does not change the result.

$$(a+b)+c = a+(b+c)$$

Example:

$$(2 + 3) + 5 = 2 + (3 + 5)$$
$$5 + 5 = 2 + 8$$
$$10 = 10$$

3. *Commutativity*. The order of the addends does not change the process of addition.

$$a + b = b + a$$

Example:

$$10 + 2 = 2 + 10$$
$$12 = 12$$

4. *Additive identity*. The number 0 is called the "identity element" of addition because every number added to it gives the same number.

$$a + 0 = a$$

ARITHMETIC
Fundamentals of Arithmetic

Example:

$$10 + 0 = 10$$

c) Subtraction of natural numbers
Subtraction is the mathematical operation that is
signified by the minus sign $(-)$

$$a - b = c$$

where

a = minuend
b = subtrahend
c = difference

Properties of the subtraction of natural numbers:

1. *No closure*. The result of subtraction is not always
another natural number.

$$5 - 7 \notin N$$

2. *Not commutative*.

$$10 - 3 \neq 3 - 10$$

d) Multiplication of natural numbers

Multiplication is the mathematical operation that is
signified by a multiplication cross or dot (x or ·). To
multiply two natural numbers means to add one factor

to itself as many times as are indicated by the other factor.

$$a \cdot b = c$$

where
a and b = factors
c = product

Properties of the multiplication of natural numbers include the following:

1. *Closure*. The result of multiplying two natural numbers is another natural number

$$a \cdot b \in N$$

2. *Associativity*. The way in which the factors are grouped does not change the result.

$$(a \cdot b) \cdot c = a \cdot (b \cdot c)$$

Example:

$$(3 \times 6) \times 2 = 3 \times (6 \times 2)$$
$$18 \times 2 = 3 \times 12$$
$$36 = 36$$

ARITHMETIC
Fundamentals of Arithmetic

---

3. *Commutativity*. The order of the factors does not change the product.

$$a \cdot b = b \cdot a$$

Example:

$$8 \times 3 = 3 \times 8$$
$$24 = 24$$

4. *Multiplicative identity*. The number 1 is the neutral factor of multiplication because any natural number multiplied by it gives the same number.

$$a \cdot 1 = a$$

Example:

$$5 \times 1 = 5$$

5. *Distributivity*. The multiplication of a natural number and a sum is equal to the sum of the product of the natural number with each of the addends.

$$a \cdot (b + c) = a \cdot b + a \cdot c$$

Example:

$$2 \times (5 + 4) = 2 \times 5 + 2 \times 4$$
$$2 \times 9 = 10 + 8$$
$$18 = 18$$

ARITHMETIC
Fundamentals of Arithmetic

6. *Reversibility*. The reverse of the distributive property is the extraction of a common factor.

$$a \cdot b + a \cdot c = a \cdot (b + c)$$

Example:

$$2 \text{ x } 4 + 2 \text{ x } 5 = 2 \text{ x } (4 + 5)$$
$$8 + 10 = 2 \text{ x } 9$$
$$18 = 18$$

e) Division of natural numbers.

If a, b, and c are natural numbers $(a,b,c \in N)$ and a is divided by b, with the result c, this can be written:

$$a : b = c$$

where

a = dividend
b = divisor
c = quotient

Type of division:

1. A division operation is exact when the remainder is zero.

$$a = b \cdot c$$

Example:

$$20 : 5 = 4$$
$$\underline{-20}$$
$$0$$

ARITHMETIC
Fundamentals of Arithmetic

---

checking: $20 = 5 \cdot 4$

   2.   A division is not exact when the remainder is not zero.

$$a = b \cdot c + r$$

where

   r = remainder

Example:

$$17 : 5 = 3$$
$$\frac{-15}{2}$$

checking: $17 = 5 \cdot 3 + 2$

Properties of the division of natural numbers:

1. *No closure*. The result of dividing two natural numbers is not always another natural number.

Example:

$$3 : 5 \notin N$$

2. *No commutativity*.

$$a : b \neq b : a$$

Example:

$$8 : 2 \neq 2 : 8$$

ARITHMETIC
Fundamentals of Arithmetic

---

3. *Zero divided by a number equals zero*.

$$0 : a = 0$$

Example:

$$0 : 4 = 0$$

4. *Division by* 0 *is undefined*.

f) Order of performing operations with natural numbers.

Mathematicians have devised a standard order of operations for calculations involving more than one arithmetical operation:

Rule 1: First, perform any calculations inside parentheses.

Rule 2: Next, perform all multiplications and divisions, working from left to right.

Rule 3: Lastly, perform all additions and subtractions, working from left to right.

Example 1:

$$3 + 5 \times 7 = ?$$
$$3 + 5 \times 7 = 3 + 35 = 38$$

ARITHMETIC
Fundamentals of Arithmetic

Example 2:

$$(1 + 3) \times (8 - 4) = ?$$
$$(1 + 3) \times (8 - 4) = 4 \times 4 = 16$$

2. Integers

An integer is a whole number that has no fractional part and no digits after the decimal point. An integer can be positive, negative, or zero. A set of integers, denoted by *Z,* is defined as follows:

$$Z = \left\{ ...,-3, -2, -1, 0, 1, 2, 3,... \right\}$$

The set *Z* is a *denumerable* set.
The set of integers consisting of zero (0) and the natural numbers (1, 2, 3,...) are also called the *whole numbers*.
The set of integers consisting of -1, -2, -3,... is called the negative integers.

Example:

325, 483, -65, 99 are integers,
but 99.3 and 0.25 are not

ARITHMETIC
Fundamentals of Arithmetic

---

3. Divisibility of Numbers

The property of divisibility of numbers provides a quick way to find the factors of large numbers.

a) Numbers divisible by 2

Numbers are divisible by 2 if the ones digit is evenly divisible by 2. This means that even numbers are divisible by 2.

b) Numbers divisible by 3

Numbers are divisible by 3 if the sum of all the individual digits is evenly divisible by 3.

Example:

The number 3630 is divisible by 3 because the sum of all the individual digits is 12, which is evenly divisible by 3.

c) Numbers divisible by 4

Numbers are divisible by 4 if the number formed by the last two individual digits is evenly divisible by 4.

Example:

The number 3624 is divisible by 4 because 24 is evenly divisible by 4.

ARITHMETIC
Fundamentals of Arithmetic

d) Numbers divisible by 5

Numbers are evenly divisible by 5 if the last digit of the number is 0 or 5.

e) Numbers divisible by 6

Numbers are evenly divisible by 6 if they are evenly divisible by both 2 and 3.

Example:

Number 3636 is divisible by 6 because sum of all individual digits is 18, which is evenly divisible by 3, and the last digit (6) is an even number that is evenly divisible by 2.

f) Numbers divisible by 7

To determine whether a number is divisible by 7, take the last digit off the number, double it and subtract the doubled number from the remaining number. If the result is a multiple of 7 (e.g. 14, 7, −7, etc.), then the number is evenly divisible by 7. This may need to be repeated several times.

Example question:

Is the number 3101 evenly divisible by 7?

Solution:

Step 1: Take the last digit (1) off of the number

ARITHMETIC
Fundamentals of Arithmetic

Step 2: Double the removed digit and subtract it from the remaining numbers

$$
\begin{array}{r}
310 \\
-\ 2 \\
\hline
308
\end{array}
$$

Step 3: Repeat the process by taking off the 8
Step 4: Double the number 8 to get 16 and subtract it from the remaining numbers

$$
\begin{array}{r}
30 \\
-16 \\
\hline
14
\end{array}
$$

The result is 14, which is a multiple of 7, so the number 3101 is evenly divisible by 7.

　g) Numbers divisible by 8
Numbers are divisible by 8 if the number formed by the last three individual digits is evenly divisible by 8.

Example:
The number 3632 is evenly divisible by 8 because the last three digits form a number, 632, which is evenly divisible by 8.

ARITHMETIC
Fundamentals of Arithmetic

---

h) Numbers divisible by 9

Numbers are divisible by 9 if the sum of all the individual digits is evenly divisible by 9.

Example:

Number 369 is evenly divisible by 9 because the sum of the digits of number 369 is 18, which is evenly divisible by 9.

i) Numbers divisible by 10

Numbers are divisible by 10 if the last digit is a 0.

4.   Prime and Composite Numbers

Prime numbers. A prime number is a positive integer greater than 1 that can be divided evenly only by 1 or itself.

Example:

The number 7 is a prime number because it can be divided evenly only by 1 or 7.

Composite numbers. A composite number is a positive integer bigger than 1, which can be divided evenly by numbers other than 1 or itself. The first few composite numbers are 4, 6, 8, 9, 10, 12, 14, 15, 16, 18,...

ARITHMETIC
Fundamentals of Arithmetic

There is an infinite number of composite numbers. The number 1 is a special case that is considered to be neither composite nor prime.

5. Decomposing Numbers into their Prime Factors

Every composite number can be decomposed into simple factors. Decomposing a composite number into prime factors allows us to write this number as a product of prime numbers.

Example 1

$$18 = 2 \cdot 3 \cdot 3; \ 8 = 2 \cdot 2 \cdot 2$$

Always start with dividing by the smallest prime number and dismantle until the quotient cannot be further dismantled. Then divide by the next prime number and repeat several times until no longer possible.

Example 2:

$$80 \mid 2$$
$$40 \mid 2$$
$$20 \mid 2$$
$$10 \mid 2$$
$$5 \mid 5$$
$$1$$

5 is a prime number, so it cannot be dismantled further.
Hence, $80 = 2 \cdot 2 \cdot 2 \cdot 2 \cdot 5$

6. Greatest Common Divisor (GCD)

The largest positive integer that divides two or more
integers without any remainder is called the Greatest
Common Divisor or the Greatest Common Factor
(GCF). One good way to find the GCD is to write
down all the factors of all numbers and then find the
biggest one that appears in all the lists.

Example 1:
Find the GCD of 12 and 8.

Solution:

Factors of 12: 1 , 2 , 3 , **4** , 6 , 12
Factors of 8: 1 , 2 , **4** , 8

The largest divisor they both have is 4.

Hence, the GCD (12, 8) = 4

Example 2:
Disassembly of composite numbers 22 and 36 into
prime factors:

ARITHMETIC
Fundamentals of Arithmetic

$$22| : 2 \qquad 36| : 2$$
$$11| : 11 \qquad 18| : 2$$
$$1 \qquad\qquad 9| : 3$$
$$3| : 3$$
$$1$$

So the prime factors are:

$$22 = 2 \cdot 11$$
$$36 = 2 \cdot 2 \cdot 3 \cdot 3$$

The largest common divisor they both have is 2.
Hence, the GCD (22, 36) = 2

7. Least Common Multiple (LCM)

The least common multiple (LCM) of two numbers is
the smallest number into which they both divide
evenly. One good way to find the least common
multiple of two numbers is to multiply both numbers
by 1,2,3,4,5... and then find the first multiple that
appears in both lists.

Example 1:
Find the least common multiple of 6 and 8.

ARITHMETIC
Fundamentals of Arithmetic

Solution:

Multiples of 6:   6, 12, 18, **24**, 30, ...
Multiples of 8:   8, 16, **24**, 32, 40, ...

The first number that appears in both lists is 24, so the LCM (6,8) = 24.

To find the LCM of two or more numbers without listing all of the multiples, use the prime factorization

Example 2:
What is the LCM of 9 and 12?

Solution:

Step 1: Factor both numbers to their prime factors

$$
\begin{array}{rl}
9 & 3 \\
3 & 3 \\
1 &
\end{array}
\qquad
\begin{array}{rl}
12 & 3 \\
4 & 2 \\
2 & 2 \\
1 &
\end{array}
$$

Step 2: Identify any shared primes. With 9 and 12, 3 is a shared prime number.

ARITHMETIC
Fractions

Step 3: Take the shared prime and multiply it with all of the other prime factors.

$$\mathbf{3} \times 3 \times 2 \times 2 = 36$$

The bold 3 is the shared prime factor.
The rest of the numbers are the other prime factors.

8. Fractions

A fraction, also called a common fraction, is a way of expressing a number that is a ratio of two integers. It is written in the following form:

$$\frac{a}{b}, \ b \neq 0$$

where

a = numerator—the number of parts to count
b = denominator—the number of equal parts into which the number 1 has been divided

1. Proper fraction: $\dfrac{a}{b} < 1, \ b \neq 0$

2. Mixed number: $c\dfrac{a}{b}, \ b \neq 0$

ARITHMETIC
Fractions

---

where
c = whole number
$\dfrac{a}{b}$ = proper fraction

3. Improper fraction: $\dfrac{a}{b} \geq 1, \quad b \neq 0$

4. Changing a mixed number to an improper fraction:

$$c\frac{a}{b} = \frac{(b \cdot c) + a}{b}, \quad b \neq 0$$

5. Decimal fraction:
   This is a fraction whose denominator is not written but is understood to be a power of 10.

Examples:

$$0.9 = \frac{9}{10}; \quad 0.54 = \frac{54}{100}; \quad 0.123 = \frac{123}{1000}$$

A general method for changing a fraction to a decimal is to divide the numerator by the denominator.

ARITHMETIC
Fractions

Example:

$$\frac{5}{8} = 0.625$$

6. Equality: $\dfrac{a}{b} = \dfrac{c}{d}$ if and only if $ad = bc$

7. Equivalence: $\dfrac{a}{b} \quad \dfrac{ac}{bc} \quad = \quad (c \neq 0)$

8. Addition:

To add or subtract fractions, the denominators must be the same. To add or subtract fractions with different denominators, first convert each fraction to an equivalent fraction with the same denominator, and add or subtract only the numerator, keeping the same denominators.

Example:

$$\frac{2}{5} + \frac{1}{5} = \frac{2+1}{5} = \frac{3}{5}; \quad \frac{1}{2} + \frac{1}{3} = \frac{1 \cdot 3 + 1 \cdot 2}{2 \cdot 3} = \frac{5}{6}$$

9. Addition of mixed numbers:
   Add the whole numbers first and add the separately.

ARITHMETIC
Fractions

Example:

$$4\frac{3}{8} + 2\frac{2}{8} = 6\frac{5}{8}$$

10. Multiplication:
   To multiply a fraction by a fraction, multiply the numerators and multiply the denominators.

Example:

$$\frac{2}{3} \cdot \frac{5}{7} = \frac{10}{21}$$

To multiply a fraction by a whole number, multiply the numerator by the whole number. Do not change the denominator.

Example:

$$5 \cdot \frac{2}{8} = \frac{10}{8} = 1\frac{2}{8}$$

11. Division:
   To divide a fraction by a fraction, divide the numerators and divide the denominators, if possible; if not, then take the reciprocal of the divisor, and multiply the fractions.

ARITHMETIC
Fractions

Example:

$$\frac{3}{8} \div \frac{1}{4} = \frac{3}{2} = 1\frac{1}{2}; \quad \frac{5}{8} \div \frac{2}{3} = \frac{5}{8} \cdot \frac{3}{2} = \frac{15}{16}$$

12. Expanding and reducing fractions:
    Expanding a fraction means that the numerator and denominator are multiplied by the same number.

Example 1:

$$\frac{2}{5} = \frac{2 \cdot 3}{5 \cdot 3} = \frac{6}{15}$$

Reducing a fraction means that the numerator and denominator are divided by the same integer.

Example 2:

$$\frac{8}{12} = \frac{8 : 4}{12 : 4} = \frac{2}{3}$$

A fraction that cannot be reduced is said to be irreducible. Some irreducible fractions:

$$\frac{2}{3}; \frac{3}{5}; \frac{5}{9}$$

ARITHMETIC
Proportionality and Percentages

---

9. Proportionality

Direct proportionality. When quantity y is increased/ reduced by a certain number of times, quantity x is increased/reduced by the same number of times.
The coefficient of proportionality between x and y is:

$$k = \frac{x}{y}, \quad k > 0$$

Example:

A tailor sews five suits for 12 days. How many days does he need to sew 23 suits?

Solution:

1st way: reducing to the unit rate (k)

5 suits	12 days
1 suit	$k = 12 : 5 = 2.4$ days
23 suits	$k \times 23 = 2.4 \times 23 = 55.2$ days

2nd way: Triple Rule

    5 suits   12 days

↑ 23 suits x days ↑

$x : 12 = 23 : 5$

$x = \dfrac{12 \cdot 23}{5} = 55.2$ days

ARITHMETIC
Proportionality and Percentages

Answer: The tailor needs 55.2 days to sew 23 suits.

Inverse proportionality. Of two numbers, one number is increased as many times as the other is reduced. The coefficient of inverse proportionality (k) is equal to x times y:

$$k = x \cdot y, \quad k > 0$$

Example:
A runner who is running at a speed of 20 km/h reaches the goal in 0.5 h. How long will it take him to reach the goal if he is running at a speed of 15 km/h?

Solution:
1st way: by using coefficient (k)

speed x	time y	coefficient k
		$k = x \cdot y$
20 km/h	0.5 h	$= 20$ km/h x 0.5 h
		$= 10$ km

$$x = 15 \text{ km/h}$$
$$y = \frac{k}{x} = \frac{10}{15} = \frac{2}{3}\text{h} = \frac{2}{3}60 = 40 \text{ min}$$

2nd way: Triple Rule

ARITHMETIC
Proportionality and Percentages

$$\downarrow 20 \ \text{km/h} \qquad 0.5 \ \text{h}$$

$$15 \ \text{km/h} \qquad y \ \uparrow$$

$$y : 0.5 = 20 : 15$$

$$y = \frac{0.5 \times 20}{15} = \frac{10}{15} = \frac{2}{3}\text{h} = 40 \ \text{min}$$

Answer: The runner needs 40 *minutes* to get to the finish line if running at a speed of 15 km/h.

10. Percentages

A percentage, denoted by p%, is a portion of a quantity expressed as a fraction with denominator 100.
The formula for percentage is

$$p = \frac{P}{S} \cdot 100$$

where
p = percentage
P = portion of S
S = base

ARITHMETIC
Proportionality and Percentages

Example:

The price of an item is increased by $15 and now stands at $140. What was the percentage of the price increase and what was the price before the increase?

Solution:

$$S + P = 140$$
$$S + 15 = 140$$
$$S = 125$$

$$p = \frac{P}{S}100 = \frac{15}{125} \cdot 100 = 12$$

Thus, the price was increased by 12%, and the price before the increase was $125 increase was $125.

a) Percentage over 100

In the event that the size of $S + P$ is known, we are talking about a percentage over 100.

If we know $(S + P)$ and the percentage p, in order to get the base (S) and amount of increase (P), use the following formulas:

$$S = \frac{(S+P)\cdot 100}{100 + p}; \quad P = \frac{(S+P)p}{100 + p}$$

ARITHMETIC
Proportionality and Percentages

Example:

The price of a product after an increase of 12% is $80. What was the price before the increase (*S*) and by how much (*P*) did it increase?

Solution:

$$S = \frac{(S+P)\cdot 100}{100+p} = \frac{8000}{112} = 71.43$$

$$P = \frac{(S+P)p}{100+p} = \frac{(80)\cdot 12}{112} = 8.57$$

b) Percentage below 100

In the event that the size of ($S - P$) is known, we are talking about a percentage below 100.

If we know $(S - P)$ and the percentage *p,* in order to get the base (*S*) and amount of increase (*P*), use the following formulas:

$$S = \frac{(S-P)\cdot 100}{100-p}; \qquad P = \frac{(S-P)\cdot p}{100-p}$$

Example:

The price of a commodity after a 12% reduction is $110. How much was the price reduced, and what was it before the reduction?

Solution:

$$S = \frac{(S - P) \cdot 100}{100 - p} = \frac{(110) \cdot 100}{100 - 12} = 125$$

$$P = \frac{(S - P) \cdot p}{100 - p} = \frac{(110) \cdot 12}{100 - 12} = 15$$

11. Calculation for Mixtures

The mathematical method for determining the proportion in which quantities of ingredients need to be mixed to obtain the required quality is called calculation for mixtures.

If only two ingredients are mixed, it is a *simple mixture calculation*, and when more than two ingredients are mixed, it is a *complex mixture calculation*.

a) Calculation for simple mixtures

If

x_i = quantity of ith connatural component, for $i = 1, 2$

ARITHMETIC
Calculation for Mixtures

---

a_i = intensity of ith connatural component, for i = 1, 2

m = average intensity characteristics of the mixture, then m is given by:

$$m = \frac{a_1 x_1 + a_2 x_2}{x_1 + x_2}$$

It is necessary to determine:

1) mixing ratio, i.e., $x_1 : x_2 = o_1 : o_2$

2) quantites x_1 and x_2, if is known that

$$x_1 + x_2 = S$$

where

S = total amount of mixture

$o_1 : o_2$ = ratio of mixture

$$x_1 : x_2 = o_1 : o_2 \text{ and}$$
$$x_1 + x_2 = S$$

This system is solved:

ARITHMETIC
Calculation for Mixtures

$$k = \frac{S}{o_1 + o_2}, \quad x_i = k \cdot o_i, \quad i = 1, 2$$

where

k = proportionality factor

A ratio of mixing can be quickly and easily solved using the following scheme:

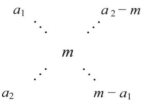

From the scheme, the mixing ratio is found:

$$x_1 : x_2 = (a_2 - m) : (m - a_1)$$

Example:
How much water should be added to dilute 2 liters of 30% strength acid to produce a 12% strength acid?

ARITHMETIC
Calculation for Mixtures

Solution:
  Scheme:

$$0 \qquad\qquad (30-12) = 18{:}6 = 3$$

$$m$$

$$30 \qquad\qquad (12-0) = 12{:}\,6 = 2$$

From the scheme, we get the ratio, $x_1 : x_2 = 3 : 2$
The problem can be solved in two ways:

1st way:

$$S = x_1 + x_2$$
$$S = x_1 + 2$$

$$k = \frac{S}{o_1 + o_2} = \frac{x_1 + 2}{3 + 2} = \frac{x_1 + 2}{5}$$

$$x_2 = k \cdot o_2 \Rightarrow k = \frac{x_2}{o_2} = \frac{2}{2} = 1$$

Then:

$$\frac{x_1 + 2}{5} = 1 \Rightarrow x_1 + 2 = 5 \Rightarrow x_1 = 3$$

ARITHMETIC
Calculation for Mixtures

2nd way:
The desired quantity is determined directly from the mixing ratio:

$$x_1 + x_2 = 3:2; \qquad x_2 = 2$$

Hence:

$$x_1 = \frac{(3) \cdot (2)}{2} = 3$$

b) Calculation for complex mixtures
A complex mixture calculation is used in situations where the mixture is composed of more than two component sizes. Such problems usually have multiple solutions. We show how to solve such a problem schematically.

Example:
We have four kinds of commodities at prices of $16, $14, $11 and $5. How should the goods be mixed to obtain 56 kg at a unit cost of $12?

Solution:
In the scheme: the left column sorts the prices by size, in the middle is the required intensity, i.e. price (12), and the right column will determine the required ratio.

ARITHMETIC
Calculation for Mixtures

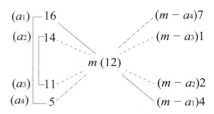

From the scheme, the ratio is:

$$a_1 : a_2 : a_3 : a_4 = 7 : 1 : 2 : 4$$

Using the proportionality constant, we get:

$$7k + k + 2k + 4k = 56$$
$$14k = 56 \Rightarrow k = 4$$

The commodities should be mixed in the following way:

- commodity at $16 $(a_1) \Rightarrow 7 \times 4 = 28$ kg
- commodity at $14 $(a_2) \Rightarrow 1 \times 4 = 4$ kg
- commodity at $11 $(a_3) \Rightarrow 2 \times 4 = 8$ kg
- commodity at $5 $(a_4) \Rightarrow 4 \times 4 = 16$ kg

Total : 56 kg

ALGEBRA

The purpose of this collection of algebraic references is to provide a brief, clear and handy guide to the more important, formal rules of algebra and the most commonly used formulas for evaluating quantities, as well as examples of their applications for solving algebraic problems.

This section contains the following:

- Fundamentals of Algebra
- Linear Equations
  Determinants
- Quadratic Equations
- Inequalities
- Sequences and Series
- Functions and Their
  Graphs

ALGEBRA
Fundamentals of Algebra

1. Sets of Real Numbers

The set of all rational numbers combined with the set of all irrational numbers gives us the set of real numbers. The relationships among the various sets of real numbers are shown below.

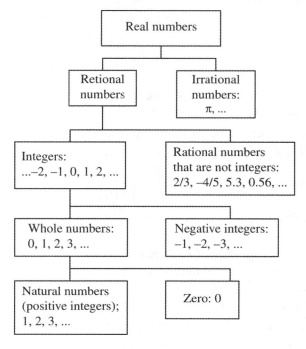

ALGEBRA
Fundamentals of Algebra

---

2. Properties of Real Numbers
If a, b, and c are real numbers, then

a) Addition properties

Commutative:	$a + b = b + a$
Associative:	$(a + b) + c = a + (b + c)$
Identity:	$a + 0 = 0 + a = a$
Inverse:	$a + (-a) = (-a) + a = 0$

b) Multiplication properties

Commutative:	$ab = ba$
Associative:	$(ab)c = a(bc)$
Identity:	$a \cdot 1 = 1 \cdot a = a$
Inverse:	$a\left(\dfrac{1}{a}\right) = \left(\dfrac{1}{a}\right)a = 1 \ (a \neq 0)$
Distributive:	$a(b + c) = ab + ac$

3. Properties of Equality
If a, b, and c are real numbers, then

Identity:	$a = a$
Symmetric:	If $a = b$, then $b = a$
Transitive:	If $a = b$ and $b = c$, then $a = c$
Substitution:	If $a = b$, then a may be replaced by b

ALGEBRA
Fundamentals of Algebra

4. Properties of Fractions

If $\dfrac{a}{b}$ and $\dfrac{c}{d}$ are fractions of real numbers, where $b \neq 0$ and $d \neq 0$, then

Equality: $\qquad \dfrac{a}{b} = \dfrac{c}{d}$ if and only if $ad = bc$

Equivalence: $\qquad \dfrac{a}{b} = \dfrac{ac}{bc} \quad (c \neq 0)$

Addition: $\qquad \dfrac{a}{b} + \dfrac{c}{b} = \dfrac{a+c}{b}$

Subtraction: $\qquad \dfrac{a}{b} - \dfrac{c}{b} = \dfrac{a-c}{b}$

Multiplication: $\qquad \dfrac{a}{b} \cdot \dfrac{c}{d} = \dfrac{ac}{bd}$

Division: $\qquad \dfrac{a}{b} \div \dfrac{c}{d} = \dfrac{a}{b} \cdot \dfrac{d}{c} = \dfrac{ad}{bc} \quad (c \neq 0)$

Sign: $\qquad -\dfrac{a}{b} = \dfrac{-a}{b} = \dfrac{a}{-b}$

$$-\left(\dfrac{-a}{b}\right) = \dfrac{a}{b}$$

ALGEBRA
Fundamentals of Algebra

---

5. Division Properties of Zero
If a is a real number, where $a \neq 0$, then

$$\frac{0}{a} = 0$$

(zero divided by any nonzero number is zero).

$$\frac{a}{0} \text{ is undefined}$$

(division by zero is undefined).

$$\frac{0}{0} \text{ is indeterminate.}$$

Expressions of this kind, in which there could be any number of values, are called "indeterminate."

6. Real Number Line
The real numbers can be represented by a real number line as shown below.

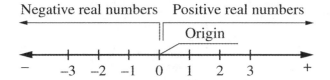

ALGEBRA
Fundamentals of Algebra

Certain order relationships exist among real numbers.
If a and b are real numbers, then

$$a = b, \qquad \text{if } a - b = 0$$

$$a > b, \qquad \text{if } a - b \text{ is positive}$$

$$a < b, \qquad \text{if } b - a \text{ is positive}$$

The symbols that represent inequality are > (greater than) and < (less than).

7. Intervals
In general, there are four type of intervals.

a) Open interval
Represents all real numbers between a and b, not including a and not including b. The interval notation is

$$(a, b)$$

b) Closed interval
Represents all real numbers between a and b, including a and including b. The interval notation is

$$[a, b]$$

ALGEBRA
Fundamentals of Algebra

c) Half-open interval

Represents all real numbers between a and b, not including a but including b. The interval notation is

$$(a,b]$$

d) Half-closed interval

Represents all real numbers between a and b, including a but not including b. The interval notation is

$$[a,b)$$

8. Absolute Value

The absolute value of the real number a, denoted by $|a|$, is defined by

$$|a| = \begin{cases} a & \text{if } a \geq 0 \\ -a & \text{if } a < 0 \end{cases}$$

a) Properties of absolute value

For all real numbers a and b,

Product: $\qquad |ab| = |a| \cdot |b|$

Quotient: $\qquad \left|\dfrac{a}{b}\right| = \dfrac{|a|}{|b|} \ (b \neq 0)$

Difference: $\qquad |a-b| = |b-a|$

Inequality:
$$|a + b| \leq |a| + |b|$$
$$|-a| = |a|$$

If $|a| = b$, then $\quad a = b$ or $a = -b$ a

If $|<|b$, then $\quad -b < a < b$

If $|a| > b$, then $\quad a > b$ or $a < -b$

9. Distance between Two Points on the Number Line

For any real numbers a and b, the distance between a and b, denoted by $d(a, b)$, is

$$d(a,b) = |a - b| \text{ , or equivalently, } |b-a|$$

10. Definition of Positive Integer Exponents

For any positive integer n, if b is any real number, then,

$$b^n = b \cdot b \cdot b \cdot \ldots b \qquad (n \text{ factors of } b)$$

where
b = the base
n = the exponent

ALGEBRA
Fundamentals of Algebra

---

11. Definition of b^0
For any nonzero real number b,

$$b^0 = 1$$

12. Definition of b^{-n}
For any natural number n,

$$b^{-n} = \frac{1}{b^n} \text{ and } \frac{1}{b^{-n}} = b^n \quad (b \neq 0)$$

Note: The expressions 0^0, 0^n where n is a negative integer, and $\dfrac{x}{0}$ are all undefined expressions.

13. Properties of Exponents
If m, n, and p are integers and a and b are real numbers, then

Product:
$$a^m a^n = a^{m+n}$$
$$a^m \cdot b^m = (ab)^m$$

Quotient:
$$\frac{a^m}{a^n} = a^{m-n} \quad (a \neq 0)$$

$$\frac{a^m}{b^m} = \left(\frac{a}{b}\right)^m \quad (b \neq 0)$$

Power:
$$\left(a^m\right)^n = a^{mn}$$

$$\left(a^m b^n\right)^p = a^{mp} b^{np}$$

$$\left(\frac{a^m}{b^n}\right)^p = \frac{a^{mp}}{b^{np}} \quad (b \neq 0)$$

14. Definition of $\sqrt[n]{a}$

The symbol $\sqrt[n]{a}$ is called a radical. $\sqrt{}$ is the radical sign, n is the index or root (which is omitted when it is 2), and a is the radicand.

$$\sqrt[n]{a} = a^{\frac{1}{n}}$$

15. Properties of Radicals

If m and n are natural numbers greater than or equal to 2, and a and b are nonnegative real numbers, then

Product:
$$\sqrt[n]{a} \cdot \sqrt[n]{b} = \sqrt[n]{ab}$$

ALGEBRA
Fundamentals of Algebra

Quotient:
$$\frac{\sqrt[n]{a}}{\sqrt[n]{b}} = \sqrt[n]{\frac{a}{b}} = \left(\frac{a}{b}\right)^{\frac{1}{n}} \quad (b \neq 0)$$

Index:
$$\sqrt[n]{\sqrt[m]{a}} = \sqrt[n \cdot m]{a}$$

$$\left(\sqrt[n]{a}\right)^m = a^{\frac{m}{n}}$$

$$\sqrt[n]{a^n} = a$$

$$\left(\sqrt[n]{a}\right)^n = a$$

16. General Form of a Polynomial

The general form of a polynomial of degree n in the variable x is

$$a_n x^n + a_{n-1} x^{n-1} + \ldots + a_1 x + a_0$$

Note:

n is a non-negative integer and $a_n \neq 0$.

The coefficient a_n is the leading coefficient, and

a_0 is the constant term.

17. Factoring Polynomials

Factoring a polynomial is writing a polynomial as a product of polynomials of lower degree.

---

a) The square of a binomial:

$$(a \pm b)^2 = a^2 \pm 2ab + b^2$$

b) The cube of a binomial:

$$(a \pm b)^3 = a^3 \pm 3a^2b + 3ab^2 \pm b^3$$

c) The difference of two squares:

$$a^2 - b^2 = (a+b)(a-b)$$

d) The sum or difference of two cubes:

$$a^3 \pm b^3 = (a \pm b)(a^2 \mp ab + b^2)$$

e) The square of a trinomial:

$$(a \pm b + c)^2 = a^2 \pm 2ab + 2ac + b^2 \pm 2$$

18. Order of Operations

If grouping symbols are present, evaluate by performing the operations within the grouping symbols first, while observing the order given in Steps 1 to 3. For example,

ALGEBRA
Fundamentals of Algebra

$$2x^3 - \left\{ x^2 - [x - (2x - 1)] + 4 \right\}$$

Step 1: Remove parentheses

$$= 2x^3 - \left\{ x^2 - [x - 2x + 1] + 4 \right\}$$
$$= 2x^3 - \left\{ x^2 - [-x + 1] + 4 \right\}$$

Step 2: Remove brackets

$$= 2x^3 - \left\{ x^2 + x - 1 + 4 \right\}$$
$$= 2x^3 - \left\{ x^2 + x + 3 \right\}$$

Step 3: Remove braces

$$= 2x^3 - x^2 - x - 3$$

The operations of multiplication and division take precedence over addition and subtraction.

19. Adding and Subtracting Polynomials

Adding:
$$\left(ax^2 + bx - c \right) + \left(a_1 x^2 - b_1 x - c \right)$$
$$= \left(a + a_1 \right) x^2 + \left(b - b_1 \right) x - 2c$$

Subtracting: $(ax^2 + bx - c) - (a_1x^2 - b_1x - c) =$
$$(a - a_1)x^2 + (b + b_1)x$$

20. Multiplying Polynomials

$(ax^2 - bx + c) \cdot (a_1x^2 - 1) =$
$ax^2(a_1x^2 - 1) - bx(a_1x^2 - 1) + c(a_1x^2 - 1) =$
$aa_1x^4 - a_1bx^3 - (a - a_1c)x^2 + bx - c$

21. Dividing Polynomials

Examples:

a) Divide poly nomial $(x^2 - 9x + 10)$ by polynomial $(x + 1)$

b) Divide polynomial $(ax^5 + bx^3 - c)$ by monomial a_1x

Solution:

a) Dividing a polynomial by a polynomial:
$$(x^2 - 9x - 10) \div (x + 1) = x - 10$$

$x^2 + x$
— — (changed sign)

ALGEBRA
Fundamentals of Algebra

$$-10x - 10$$
$$-10x - 10$$
$$\underline{+\quad\quad+\quad\text{(changed sign)}}$$
$$0$$

b) Dividing a polynomial by a monomial:

$$(ax^5 + bx^3 - c) \div a_1 x = \frac{ax^5}{a_1 x} + \frac{bx^3}{a_1 x} - \frac{c}{a_1 x} =$$

$$\frac{a}{a_1}x^4 + \frac{b}{a_1}x^2 - \frac{c}{a_1 x}$$

22. Rational Expressions

A rational expression is a fraction in which the numerator and denominator are polynomials.

For example:

$$\frac{x^2 - 4x - 21}{x^2 - 9}, \text{ or } \frac{p}{q}$$

a) Properties of rational expressions

Let $\frac{p}{q}$ and $\frac{r}{s}$ be rational expressions, where

ALGEBRA
Fundamentals of Algebra

$q \neq 0$ and $s \neq 0$

Equality: $\dfrac{p}{q} = \dfrac{r}{s}$ if and only if $ps = qr$

Equivalent expressions: $\dfrac{p}{q} = \dfrac{pr}{qr},\ r \neq 0$

Sign: $-\dfrac{p}{q} = \dfrac{-p}{q} = \dfrac{p}{-q}$

b) Operations with rational expressions

For all rational expressions $\dfrac{p}{q}$ and $\dfrac{r}{s}$, where

$q \neq 0$ and $s \neq 0$

Addition: $\dfrac{p}{q} + \dfrac{r}{q} = \dfrac{p+r}{q}$

Subtraction: $\dfrac{p}{q} - \dfrac{r}{q} = \dfrac{p-r}{q}$

Multiplication: $\dfrac{p}{q} \cdot \dfrac{r}{s} = \dfrac{pr}{qs}$

ALGEBRA
Fundamentals of Algebra

Division: $\dfrac{p}{q} \div \dfrac{r}{s} = \dfrac{ps}{qr}$, $\qquad r \neq 0$

c) Least common denominator (LCD)
To add and subtract rational expressions when the denominators are different; we must find equivalent rational expressions that have a common denominator. It is most efficient to find the LCD of the expressions:

Step 1: Factor each denominator completely and express repeated factors using exponential notation.

Step 2: Identify the largest power of each factoring in any of the factorizations. The LCD is the product of each factor raised to the largest power.

Example:
Find the LCD and add the rational expressions

$$\frac{3}{x^2 + x} \text{ and } \frac{2}{x^2 - 1}$$

Solution:

Step 1: $\qquad\qquad x^2 + x = x(x + 1)$, and

$\qquad\qquad\qquad x^2 - 1 = (x + 1)(x - 1)$

Step 2: The LCD of the two expressions is
$$x(x+1)(x-1)$$

For adding fractions, we express each fraction using the common denominator, and then we add the numerators.

$$\frac{3}{x^2+x} + \frac{2}{x^2-1} = \frac{3}{x(x+1)} + \frac{2}{(x+1)(x-1)}$$
$$= \frac{3(x-1)+2x}{x(x+1)(x-1)} = \frac{5x-3}{x(x^2-1)}$$

23. Complex Fractions
A complex fraction is a fraction whose numerator or denominator or both contain more fractions.
To simplify a complex fraction, use one of two methods:

Method 1: Find the LCD of all the denominators within the complex fraction. Then multiply both the numerator and denominator of the complex fraction by the LCD.

Method 2: First add or subtract, if necessary, to get a single fraction in both the numerator and the denominator. Then divide by multiplying by the reciprocal of the denominator.

ALGEBRA
Fundamentals of Algebra

Example: Simplify a complex fraction $\dfrac{3-\dfrac{1}{a}}{1+\dfrac{4}{a}}$

Solution:

$$\frac{3-\dfrac{1}{a}}{1+\dfrac{4}{a}} = \frac{\dfrac{3a-1}{a}}{\dfrac{a+4}{a}} = \frac{(3a-1)a}{(a+4)a} = \frac{3a-1}{a+4}$$

24. Definition of a Complex Number

A complex number is any number that can be written
$$z = a + bi$$

where

a = real part of the complex number

b = real number of imaginary part of the complex number

i = imaginary unit $\left(i = \sqrt{-1}\right)$

a) Operations with complex numbers

Let $a + bi$ and $c + di$ be complex numbers, then

Addition: $(a+bi)+(c+di)=(a+c)+(b+d)\cdot i$

ALGEBRA
Fundamentals of Algebra

Subtraction: $(a + bi) - (c + di) = (a - c) + (b - d) \cdot i$

Multiplication: $(a + bi) \cdot (c + di) =$
$(ac - bd) + (ad + bc) \cdot i$

Division: $\dfrac{a + bi}{c + di} = \dfrac{ac + bd}{c^2 + d^2} + \dfrac{bc - ad}{c^2 + d^2} i$ $(c + di \neq 0)$

b) Conjugate of a complex number

The conjugate of a complex number $z = a + bi$ is

$$\bar{z} = a - bi$$

Properties: $z + \bar{z}$ is a real number

$z \cdot \bar{z} = |z|^2$ is always a real number

$\bar{z} = z$ if and only if z is a real number

$\bar{z}^n = (z)^n$ for all natural numbers n

c) Powers of i

If n is a positive integer, then

$$i^n = i^r$$

where

r = remainder of the division of n by 4

ALGEBRA
Linear Equations

Example: Evaluate i^{37}

Use the theorem on powers of i

$i^{37} = i^1 = i$ (the remainder of $37 \div 4$ is 1)

25. Definition of a Linear Equation

An equation is a statement of equality between two mathematical expressions.

A linear equation in the single variable x can be written in the form

$$ax + b = 0$$

where

a, b = real numbers $(a \neq 0)$

26. Addition and Multiplication Properties of Equality

If	$a = b$, then	$a + c = b + c$
If	$a = b$, then	$ac = bc$
If	$-a = b$, then	$a = -b$
If	$x + a = b$, then	$x = b - a$
If	$x - a = b$, then	$x = a + b$
If	$ax = b$, then	$x = \dfrac{b}{a}$
If	$\dfrac{x}{a} = b$, then	$x = ab$

27. Systems of Linear Equations

A system of linear equations can be solved in various different ways, such as by using substitution, elimination, determinants, matrices, graphs, etc.

a) The method of substitution:

$$x + 2y = 4 \qquad (1)$$
$$3x - 2y = 4 \qquad (2)$$

The method of substitution involves five steps:

Step 1: Solve for y in equation (1)

$$y = \frac{4 - x}{2}$$

Step 2: Substitute this expression for y in equation (2). This will change equation (2) to an equation with just one variable, x

$$3x - 2\frac{4 - x}{2} = 4$$

Step 3: Solve for x in the translated equation (2)

$$4x = 8$$
$$x = 2$$

ALGEBRA
Linear Equations

Step 4: Substitute this value of x in the y equation obtained in Step 1

$$2 + 2y = 4$$
$$y = 1$$

Step 5: Check answers by substituting the values of x and y into each of the original equations. If, after the substitution, the left side of the equation equals the right side of the equation, the answers are correct.

b) The method of elimination:

$$x + 2y = 4 \qquad (1)$$
$$3x - 2y = 4 \qquad (2)$$

The process of elimination involves four steps:

Step 1: Change equation (1) by multiplying it by (-3) to obtain a new and equivalent equation (1).

$$-3x - 6y = -12, \quad \text{new equation (1).}$$

Step 2: Add new equation (1) to equation (2) to obtain equation (3).

$$-3x - 6y = -12$$
$$3x - 2y = 4$$
$$\overline{}$$
$$-8y = -8 \qquad (3)$$
$$y = 1$$

Step 3: Substitute $y = 1$ in equation (1) and solve for x.

$$x + 2 \cdot 1 = 4$$
$$x = 2$$

Step 4: Check your answers in equation (2).

$$3 \cdot 2 - 2 \cdot 1 = 4$$
$$4 = 4$$

28. Determinants

Let system (1) be

$$a_{11}x + a_{12}y = r_1$$
$$a_{21}x + a_{22}y = r_2 \qquad (1)$$

This can be used to represent any system of linear equations` The second order determinant of system (1) is

ALGEBRA
Determinants

$$D = \begin{vmatrix} a_{11} & a_{12} \\ a_{21} & a_{22} \end{vmatrix} = a_{11} \cdot a_{22} - a_{21} \cdot a_{12}$$

$$- \qquad +$$

To solve for x, insert column r in place of column x into determinant D, to get

$$D_x = \begin{vmatrix} r_1 & a_{12} \\ r_2 & a_{22} \end{vmatrix} = r_1 \cdot a_{22} - r_2 \cdot a_{12}$$

$$- \qquad +$$

$$x = \frac{D_x}{D} \quad (D \neq 0)$$

To solve for y, insert column r in place of column y into determinant D, to get

$$D_y = \begin{vmatrix} a_{11} & r_1 \\ a_{21} & r_2 \end{vmatrix} = a_{11} \cdot r_2 - a_{21} \cdot r_1$$

$$- \qquad +$$

$$y = \frac{D_y}{D} \quad (D \neq 0)$$

Example:

Solve this system of equations by determinants:

$$2x + 4y = 8$$
$$3x - 2y = 4$$

Solution:
Determinant for the system of equations is

$$D = \begin{vmatrix} 2 & 4 \\ 3 & (-2) \end{vmatrix} = 2 \cdot (-2) - 3 \cdot 4 = -16$$

Determinant for x is

$$D_x = \begin{vmatrix} 8 & 4 \\ 4 & (-2) \end{vmatrix} = 8 \cdot (-2) - 4 \cdot 4 = -32$$

$$x = \frac{D_x}{D} = \frac{-32}{-16} = 2$$

Determinant for y is

$$D_y = \begin{vmatrix} 2 & 8 \\ 3 & 4 \end{vmatrix} = 2 \cdot 4 - 3 \cdot 8 = -16$$

$$y = \frac{D_y}{D} = \frac{-16}{-16} = 1$$

29. Quadratic Equations

The standard form of quadratic equations is

$$ax^2 + bx + c = 0$$

ALGEBRA
Quadratic Equations

---

where

a, b, c = constants $\quad (a \neq 0)$

a) Solving quadratic equations by factoring.
$x^2 - 3x + 2 = 0$ is a quadratic equation in standard form; then

$$x^2 - 3x + 2 = x^2 - 2x - x + 2 = 0$$
$$(x - 2)(x - 1) = 0$$

The roots of the equation are:

$$(x - 2) = 0$$
$$x = 2,$$

and

$$(x - 1) = 0$$
$$x = 1$$

b) Solving quadratic equations using Vieta's rule. Normal form of quadratic equation:

$$x^2 + px + q = 0$$

Solutions:

$$x_{1,2} = -\frac{p}{2} \pm \sqrt{\frac{p^2}{4} - q}$$

ALGEBRA
Quadratic Equations

Vieta's rule:

$$p = -(x_1 + x_2)$$
$$q = x_1 \cdot x_2$$

c) Solving quadratic equations by completing the square.

Start with the standard form of quadratic equations:

$$ax^2 + bx + c = 0$$

Step 1: Write the equation in the form

$$x^2 + \frac{b}{a}x = -\frac{c}{a}$$

Step 2: Square half of the coefficient of x.

Step 3: Add the number obtained in step 2 to both sides
of the equation in step 1, factor, and solve for x.

Example:
Solve the quadratic equation by completing the square:

$$x^2 - 2x - 2 = 0$$

Solution:

Step 1: $\qquad\qquad x^2 - 2x = 2$

ALGEBRA
Quadratic Equations

Step 2:

$$\left(-\frac{2}{2}\right)^2 = 1$$

Step 3:

$$x^2 - 2x + 1 = 2 + 1$$
$$(x-1)^2 = 3$$
$$x_{1,2} = 1 \pm \sqrt{3}$$
$$x_1 = 1 + \sqrt{3}$$
$$x_2 = 1 - \sqrt{3}$$

d) Solving quadratic equations by using the quadratic formula.
The quadratic equation

$$ax^2 + bx + c = 0$$

with real coefficients, where $a \neq 0$, can be solved as follows:

$$x_{1,2} = \frac{-b \pm \sqrt{b^2 - 4ac}}{2a}$$

where

$b^2 - 4ac$ = discriminant D of the quadratic equation.

ALGEBRA
Quadratic Equations

If $D = b^2 - 4ac > 0$, then the quadratic equation has two real and distinct roots.

If $D = b^2 - 4ac = 0$, then the quadratic equation has a real root that is a double root.

If $D = b^2 - 4ac < 0$, then the quadratic equation has two distinct but no real roots.

Example:

Classify the roots of each quadratic equation:

$$1) \quad 2x^2 - 5x + 1 = 0$$
$$2) \quad 3x^2 + 6x + 7 = 0$$

Solution:

1) $D = b^2 - 4ac = (-5)^2 - 4(2)(1) = 25 - 8 = 17$

$D = 17 > 0$

Because $D > 0$, thequadratic equation $2x^2 - 5x + 1 = 0$ has two distinct real roots.

2) $D = b^2 - 4ac = (6)^2 - 4(3)(7) = 36 - 84 = -48$

$D = -48 < 0$

Because $D < 0$, the quadratic equation $3x^2 + 6x + 7 = 0$ has two distinct but no real roots.

ALGEBRA
Sequences and Series

---

30. Properties of Inequalities
For real numbers a, b, and c, the properties of inequalities are as follows:

If $a < b$, then $a + c < b + c$
   (Adding the same number to each side of an
   inequality preserves the order of the inequality.)

If $a < b$ and $c > 0$, then $ac < bc$

   (Multiplying each side of an inequality by the same
   positive number preserves the order of the
   inequality.)

If $a < b$ and $b < c$, then $a < c$

If $a < b$ and $c < d$, then $a + c < b + d$

If $0 < a < b$ and $0 < c < d$, then $ac < bd$

31. Arithmetic Sequences
The sequence 1, 4, 7, 10, …is an example of an
arithmetic sequence or arithmetic progression. The
difference between successive terms is the same
constant d. In general, an arithmetic sequence is

$$a_1, (a_1 + d), (a_1 + 2d), (a_1 + 3d),...$$

ALGEBRA
Sequences and Series

The nth term of an arithmetic sequence is

$$a_n = a_1 + (n-1)d$$

where

$d =$ common difference $\left[d = (a_n - a_{n-1}) \right]$

$a_1 =$ the first term

a) Arithmetic mean

Each term of an arithmetic sequence is the arithmetic mean of its adjacent terms:

$$a_m = \frac{a_{m-1} + a_{m+1}}{2}$$

where

$a_m =$ arithmetic mean

$a_{m-1}, \ a_{m+1} =$ adjacent terms

32. Arithmetic Series

The sum of terms in an arithmetic sequence is called an arithmetic series.

a) Sum of the first n terms

The sum of the terms of an arithmetic sequence is given by the formula

$$S_n = \frac{n}{2}(a_1 + a_n)$$

---

where
$$a_n = a_1 + (n-1)d \quad (n = 1, 2, 3, \ldots)$$

An alternative formula for the sum of an arithmetic series is

$$S_n = \frac{n[2a_1 + (n-1)d]}{2}$$

Example:
Find S_{20} for the arithmetic sequence whose first term is $a_1 = 3$ and whose common difference is $d = 5$.

Solution:
Substituting $a_1 = 3$, $d = 5$, and $n = 20$ into the formula gives

$$S_{20} = \frac{20}{2}[2(3) + (20-1)5] = 1010$$

33. Geometric Sequences

The sequence $a_1, \ a_1 r, \ a_1 r^2, a_1 r^3, \ldots, a_1(r^{n-1}), \ldots$ is called a geometric sequence. The ratio between two successive terms is the same constant r. This constant is called the common ratio.

a) The nth term of a geometric sequence is
$$a_n = a_1 r^{n-1}$$

ALGEBRA
Sequences and Series

where

$a_1 = $ the first term

$r = $ common ratio ($r = \dfrac{a_{i+1}}{a_i}$)

b) Geometric mean

Each term of a geometric sequence is the geometric mean of its adjacent terms:

$$a_m = \sqrt{a_{m-1} \cdot a_{m+1}} \qquad (1 < m < n)$$

where

$a_m = $ geometric mean

$a_{m-1}, a_{m+1} = $ adjacent terms

34. Geometric Series

The sum of items in a geometric sequence is called a geometric series.

$$S_n = a_1 + a_1 r + a_1 r^2 + ... + a_1 r^{n-2} + a_1 r^{n-1}$$

a) Sum of n terms:

$$S_n = a_1 \frac{1 - r^n}{1 - r} \qquad (r \neq 1)$$

where

r = the common ratio ($r = \dfrac{a_{i+1}}{a_i}$)

Example:
Bob saves $150 in January, and each month thereafter Bob manages to save half of what he saved the previous month. How much does Bob save in the 12th month, and what are his total savings after 12 months?

Solution:
The amounts saved each month form a geometric sequence with $a_1 = 150$, $r = 0.5$ and $n = 12$: then

$a_n = a_1 r^{n-1}$ gives

$$a_{12} = 150 \left(\frac{1}{2} \right)^{12-1} = 150 \left(\frac{1}{2} \right)^{11} = 0.073$$

This means that Bob saves 7.3 cents in the 12th month. Bob's total savings is:

$$S_{12} = 150 \frac{1 - \left(\dfrac{1}{2} \right)^{12}}{1 - \dfrac{1}{2}} = 299.9267$$

The total amount saved is $299.93

ALGEBRA
Sequences and Series

---

b) Sum of an infinite geometric series

If a_n is a geometric sequence with $|r| < 1$, $n \to \infty$ and first term a_1, then the sum of the infinite geometric series is

$$S = \frac{a_1}{1-r}$$

35. Binomial Theorem

For any binomial $a + b$ and any natural number n,

$$(a+b)^n = \binom{n}{0}a^n b^0 + \binom{n}{1}a^{n-1}b^1 + \binom{n}{2}a^{n-2}b^2 + \ldots$$

$$+ \binom{n}{n-1}a^1 b^{n-1} + \binom{n}{n}a^0 b^n$$

$$= \sum_{k=0}^{n} \binom{n}{k}a^{n-k}b^k$$

where

$$k = \text{binomial coefficient, } \binom{n}{k} = \frac{n!}{k!(n-k)!}.$$

a) A specific term of a binomial expansion

The $(k+1)$st term of the expansion of $(a+b)^n$ is given by

ALGEBRA
Sequences and Series

$$\binom{n}{k} a^{n-k} b^{k}$$

Example:

Find the fifth term in the expansion of $\left(2x^{3} - 3y^{2}\right)^{6}$

Solution:

First, we note that $5 = 4 + 1$. Thus, by inserting, $a = 2x^{3}$, $b = -3y^{2}$, $k = 4$, and $n = 6$ into the forumula, we have

$$\binom{n}{k} a^{n-k} b^{k} =$$

$$\binom{6}{4} \left(2x^{3}\right)^{2} \left(-3y\right)^{4} =$$

$$\frac{6!}{4!(6-4)!} \left(2x^{3}\right)^{2} \left(-3y^{2}\right)^{4} =$$

$$15\left(4x^{6}\right)\left(81y^{8}\right) = 4860x^{6}y^{8}$$

The fifth term is $4860x^{6}y^{8}$

---

36. The Cartesian Coordinate System

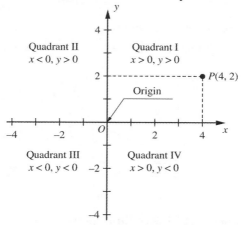

The Cartesian coordinate system in two dimensions, also known as a rectangular coordinate system, is commonly defined by two axes at right angles to each other and forming an *xy*-plane. The horizontal axis is labeled *x*, and the vertical axis is labeled *y*. The point of intersection, where the axes meet, is called the *origin* and is normally labeled *O*. To plot a point *P*(*a, b*) means to draw a dot at its location in the coordinate plane. In the figure we have plotted the point *P*(4, 2).

37. Linear Functions
A linear function is a function that can be represented by a linear equation of the form

ALGEBRA
Functions and Their Graphs

$$f(x) = y = mx + b$$

where
m and b = real constants.
The graph of function $f(x) = y = mx + b$ is

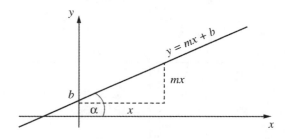

38. Forms of Linear Equations
General form: $Ax + By + C = 0$

where
A, B, C = constants ($A \neq 0, B \neq 0$)

Slope-intercept form: $y = mx + b$

where
m = slope of the line ($m = \tan\alpha$)
b = intercept on the y-axis

Vertical line: $x = a$

Horizontal line: $y = b$

Point-slope form: $y - y_1 = m(x - x_1)$

Intercept form: $\dfrac{x}{a} + \dfrac{y}{b} = 1$

Two-point form: $y - y_1 = \left(\dfrac{y_2 - y_1}{x_2 - x_1} \right)(x - x_1)$

39. Quadratic Functions

A quadratic function is a non-linear function that can be represented by an equation of the form

$$f(x) = y = ax^2 + bx + c, \qquad a \neq 0$$

where

$a, b, c =$ real numbers

The graph of $f(x) = y = ax^2 + bx + c$ is a parabola.

ALGEBRA
Functions and Their Graphs

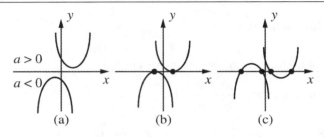

(a) (b) (c)

a) Properties of quadratic functions:

1. If $a > 0$, the parabola opens upward
2. If $a < 0$, the parabola opens downward
3. The vertex of the parabola is $\left[-\dfrac{b}{2a}, f\left(-\dfrac{b}{2a}\right)\right]$
4. The axis of symmetry of the parabola is $x = -\dfrac{b}{2a}$

5. The x-intercepts are found by solving $f(x) = 0$
6. The y-intercept is $f(0) = c$

b) Using the discriminant

When a, b, and c in equation $ax^2 + bx + c = 0$ are real numbers, then a graph of $f(x) = y = ax^2 + bx + c$ can appear in three ways:

ALGEBRA
Functions and Their Graphs

If $b^2 - 4ac < 0$, the graph of $f(x)$ does not cross the
x-axis (Fig. A)

If $b^2 - 4ac = 0$, the graph of $f(x)$ touches the x-axis at
one point (Fig. B)

If $b^2 - 4ac > 0$, the graph of $f(x)$ crosses the x-axis at
two points (Fig. C)

40. Basic Operations of Functions
If the ranges of functions f and g are subsets of the real
numbers, then

Sum: $(f + g)(x) = f(x) + g(x)$

Difference: $(f - g)(x) = f(x) - g(x)$

Product: $(f \cdot g)(x) = f(x) \cdot g(x)$

Quotient: $\left(\dfrac{f}{g}\right)(x) = \dfrac{f(x)}{g(x)}, \quad g(x) \neq 0$

41. Exponential Functions
The function defined by

$$f(x) = y = a^x \quad (a \neq 1)$$

is called an exponential function,

ALGEBRA
Functions and Their Graphs

---

where
> a = base (positive constant)
> x = exponent (any real number)

The graph of $f(x) = y = a^x$ $(a \neq 1)$ is

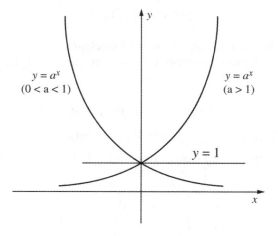

a) Properties of exponential functions

1. The domain consists of all real numbers x, $(-\infty, \infty)$.
2. The range consists of all positive numbers $(0, \infty)$.
3. The function increases when $a > 1$, and it decreases when $0 < a < 1$.
4. The graph passes through point $(0, 1)$.

ALGEBRA
Functions and Their Graphs

---

42. Natural Exponential Function

The function defined by

$$f(x) = e^x$$

is called the natural exponential function,

where
    e = base (e = 2.71828183...)
    x = exponent (any real number).

43. Logarithmic Functions
The function defined by

$$f(x) = y = \log_a x \text{ if and only if } a^y = x$$

is called a logarithmic function,

where
    a = base (a $\neq$ 1)
    x = argument (any ....   number)

   a) Properties of logarithmic functions

1. The domain consists of all positive numbers $x, (0, \infty)$.
2. The range consists of all real numbers y $(-\infty, \infty)$.

ALGEBRA
Functions and Their Graphs

---

3. The function increases from left to right if $a > 1$, and it decreases from left to right if $0 < a < 1$.
4. The graph passes through point $(1, 0)$.
5. The graph is an unbroken curve devoid of holes or breaks.

A graph of $y = \log_a x$ is

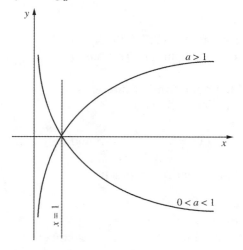

b) Laws of logarithms:

$$\log_a (x \cdot y) = \log_a x + \log_a y$$

$$\log_a \frac{x}{y} = \log_a x - \log_a y$$

$$\log_a x^n = n \log_a x$$

ALGEBRA
Functions and Their Graphs

$$\log_a \sqrt[n]{x} = \frac{1}{n}\log_a x$$

$$\log_a 1 = 0$$

$$\log_a a = 1$$

c) Notation for logarithms with special bases:

$\log x = \log_{10} x$ (common logarithm)

$\ln x \equiv \log_e x$ (natural logarithm)

GEOMETRY

Geometry is the branch of mathematics concerned with the properties of and relationships between points, lines, planes, angles, and solids, as well as with generalizations of these concepts.

If geometry has always been your nemesis, here we will explain simply and easily how to do every kind of geometrical problem you are likely to face in your profession or study of mathematics in a high school or college, from angles to solid bodies.

This section contains the most frequently used formulas, rules, and definitions regarding the following:

- Angles
- Areas
- Solid Bodies

GEOMETRY
Angles

---

1. Definition of an Angle

Two rays that share the same endpoint form an angle.
The point where the rays intersect is called the *vertex* of
the angle. The two rays are called the *sides* of the angle.

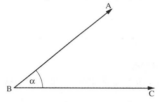

2. Unit Measurement of Angles

The radian measure of the angle φ is the ratio of the arc
length to the radius.

$$\varphi_{(\text{radian})} = 2\pi \cdot \frac{\phi^{\circ}}{360^{\circ}}$$

1 radian = 57.2957°

$$\phi^{\circ} = 360^{\circ} \cdot \frac{\varphi_{(\text{radian})}}{2\pi}$$

3. Acute Angles

An acute angle is an angle measuring between 0 and 90
degrees.

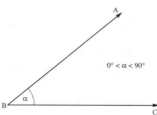

GEOMETRY
Angles

---

4. Obtuse Angles

An obtuse angle is an angle measuring between 90 and 180 degrees.

5. Right Angles

A right angle is an angle measuring exactly 90 degrees.

6. Complementary Angles

Two angles are called complementary angles if the sum of their degree measurements equals 90 degrees.

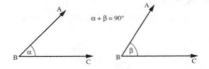

7. Supplementary Angles

Two angles are called supplementary angles if the sum of their degree measurements equals 180 degrees.

GEOMETRY
Angles

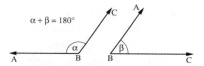

8. Vertical Angles

For any two lines that meet, such as in the diagram below, angle α and angle β are called vertical angles. Vertical angles have the same degree measurement. Angle γ and angle δ are also vertical angles.

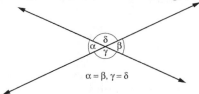

9. Alternate Interior Angles

For any pair of parallel lines 1 and 2 that are both intersected by a third line, such as line 3 in the diagram below, angle α and angle β are called alternate interior angles. Alternate interior angles have the same degree measurement. Angle γ and angle δ are also alternate interior angles.

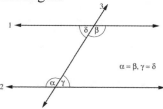

GEOMETRY
Angles

---

10. Alternate Exterior Angles

For any pair of parallel lines 1 and 2 that are both intersected by a third line, such as line 3 in the diagram below, angle α and angle β are called alternate exterior angles. Angle γ and angle δ are also alternate exterior angles.

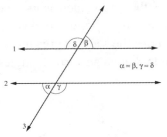

11. Corresponding Angles

For any pair of parallel lines 1 and 2 that are both intersected by a third line, such as line 3 in the diagram below, angle α and angle β are called corresponding angles. Angle γ and angle δ are also corresponding angles.

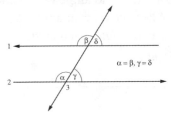

12. Angle Bisector

An angle bisector is a ray that divides an angle into two equal angles.

GEOMETRY
Angles

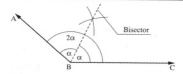

13. Perpendicular Angles

Two angles whose rays meet at right angles are perpendicular.

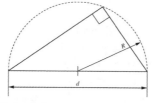

14. Thales' Theorem

A triangle inscribed in a semicircle with radius R, and diameter d is a right triangle, as shown below:

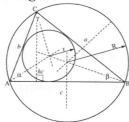

15. Oblique Triangle

GEOMETRY
Areas

An oblique triangle is any triangle that is not a right triangle. It could be an acute triangle (all three angles of the triangle are smaller than right angles) or it could be an obtuse triangle (one of the three angles is greater than a right angle).

a) Area:

$$A = \frac{c \cdot h_c}{2} = \frac{b \cdot h_b}{2} = \frac{a \cdot h_a}{2} = \sqrt{s(s-a)(s-b)(s-c)}$$

where

a, b, c = sides of the triangle

h_a, h_b, h_c = altitudes

$$s = \frac{a+b+c}{2}$$

b) Circumscribed circle

The point where the perpendicular bisectors of each side of a triangle meet is the center of the circle that circumscribes the triangle. The radius R of a circumscribed circle around a triangle is

$$R = \frac{abc}{4A}$$

c) Inscribed circle

The point where the bisectors of the three angles of a triangle meet is the center of an inscribed circle in the triangle. The radius r of an inscribed circle in a triangle is

GEOMETRY
Areas

$$r = \frac{A}{s}$$

d) Sum of the angles in a triangle:

$$\alpha + \beta + \delta = 180°$$

16. Geocenter of a Triangle

The medians of a triangle are the lines from each vertex to the midpoint of the opposite side. The medians always intersect at a single point, called the centroid or geocenter.

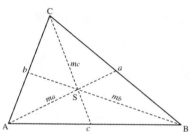

17. Orthocenter

The three altitudes of a triangle intersect at a single point, called the orthocenter of the triangle.

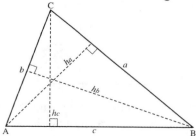

GEOMETRY
Areas

---

18. Similarity of Triangles

Two triangles are said to be similar:

1. If and only if the angles of one are equal to the corresponding angles of the other. In this case, the lengths of their corresponding sides are proportional.

2. When the triangles share an angle and the sides opposite to that angle are parallel.

3. If two angles in the different triangles are the same.

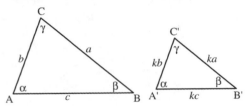

19. The Law of Cosines

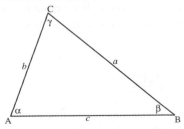

The law of cosines is valid for all triangles, even if any angle of the triangle is not a right angle.

GEOMETRY
Areas

The law of cosines can be used to compute the side lengths and angles of a triangle if all three sides or two sides and an enclosed angle are known.

$$a^2 = b^2 + c^2 - 2bc \cos \alpha$$
$$b^2 = a^2 + c^2 - 2ac \cos \beta$$
$$c^2 = a^2 + b^2 - 2ab \cos \gamma$$

20. The Law of Sines

The law of sines can be used to compute the side lengths of a triangle when two angles and one side are known. If two sides and an unenclosed angle are known, the law of sines may also be used.

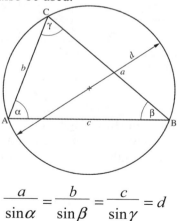

$$\frac{a}{\sin \alpha} = \frac{b}{\sin \beta} = \frac{c}{\sin \gamma} = d$$

When the last part of the equation is not used, sometimes the law is stated using the reciprocal:

GEOMETRY
Areas

$$\frac{\sin\alpha}{a} = \frac{\sin\beta}{b} = \frac{\sin\gamma}{c}$$

where

  a, b, c = sides of the triangle
  d = the diameter of the circumcircle.

21. Right Triangle

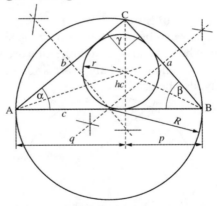

A triangle which contains a right angle (90°) is a right triangle. In the conventional a, b, c labeling of the three sides, the side of length c will represent the hypotenuse.

  a) Perimeter:
$$P = a + b + c$$

where

  a, b, c = the sides of the triangle

GEOMETRY
Areas

---

b) Area:

$$A = \frac{ab}{2} = \frac{ch_c}{2} = \frac{c}{2}\sqrt{pq} = \frac{c}{2}pq^{\frac{1}{2}}$$

where

$$a = \sqrt{pc} = (pc)^{\frac{1}{2}}$$

$$b = \sqrt{qc} = (qc)^{\frac{1}{2}}$$

$$h_c^2 = pq$$

c) Radius of inscribed circle:

$$r = \frac{ab}{a+b+c} = s - c$$

d) Radius of circumscribed circle:

$$R = \frac{c}{2}$$

22. Ratios of the Sides of a Right Triangle

A ratio is a comparison by division. Each ratio of sides of a right triangle is assigned a name, and these names are called *trigonometric functions*.

GEOMETRY
Areas

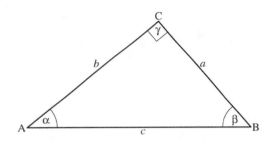

$$\sin \alpha = \frac{a}{c}, \quad \sin \beta = \frac{b}{c}$$

$$\cos \alpha = \frac{b}{c}, \quad \cos \beta = \frac{a}{c}$$

$$\tan \alpha = \frac{a}{b}, \quad \tan \beta = \frac{b}{a}$$

$$\alpha + \beta = 90°, \quad \gamma = 90°$$

23. Pythagorean Theorem

The Pythagorean theorem states that in any right triangle, the area of the square on the hypotenuse is equal to the sum of the areas of the squares on the other two sides. It can be used to find an unknown side of a right-angled triangle, or to prove that a given triangle is right- angled.

If vertex C is the right angle, we can write the theorem as

$$c^2 = a^2 + b^2$$

GEOMETRY
Areas

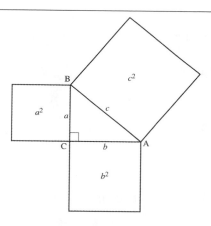

24. Equilateral Triangle

A triangle with all three sides of equal length and three $60°$ angles is an equilateral triangle.

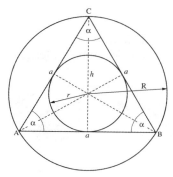

a) Angles:

$$A = B = C = \alpha = 60°$$

GEOMETRY
Areas

---

b) Perimeter:

$$P = 3a$$

c) Altitude:

$$h = \frac{a}{2}\sqrt{3}$$

d) Area:

$$A = \frac{a^2}{4}\sqrt{3}$$

e) Radius of inscribed circle:

$$r = \frac{a}{6}\sqrt{3} = \frac{R}{2}$$

f) Radius of circumscribed circle:

$$R = \frac{a}{3}\sqrt{3}$$

25. Isosceles Triangle

A triangle with two sides of equal length is an isosceles triangle.

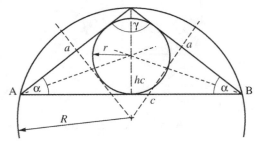

GEOMETRY
Areas

a) Area:

$$A = \frac{ch_c}{2} = \frac{c}{4}\sqrt{4a^2 - c^2}$$

b) Perimeter:

$$P = 2a + c$$

c) Altitude:

$$h_c = \frac{\sqrt{4a^2 - c^2}}{2} = a\cos\frac{\gamma}{2}$$

d) Radius of inscribed circle:

$$r = \frac{2A}{P}$$

e) Radius of circumscribed circle:

$$R = \frac{a^2 c}{4A}$$

f) Angles:

$$2\alpha + \gamma = 180°$$

where

α = base angles (congruent)

γ = vertex angle

GEOMETRY
Areas

---

26. Square

A square is a closed planar quadrilateral with all sides of equal length a, and with four right angles.

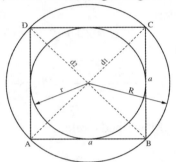

a) Perimeter:

$$P = 4a$$

b) Area:

$$A = a^2 = \frac{d^2}{2}$$

c) Radius of inscribed circle:

$$r = \frac{a}{2}$$

d) Radius of circumscribed circle:

$$R = \frac{d}{2}$$

e) Diagonals:

$$d_1 = d_2 = d = a\sqrt{2}$$

GEOMETRY
Areas

27. Rectangle

A rectangle is a closed planar quadrilateral with opposite sides of equal lengths a and b, and with four right angles.

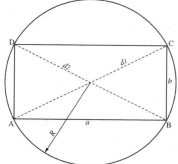

a) Perimeter:

$$P = 2(a+b)$$

b) Area:

$$A = ab$$

c) Diagonals:

$$d_1 = d_2 = d = \sqrt{a^2 + b^2}$$

d) Radius of circumscribed circle:

$$R = \frac{d}{2}$$

28. Parallelogram

A parallelogram is a closed planar quadrilateral whose opposite sides are parallel.

GEOMETRY
Areas

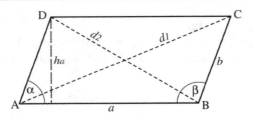

a) Perimeter:

$$P = 2(a+b)$$

b) Area:

$$A = ah_a = ab\sin\alpha$$

c) Diagonals:

$$d_1 = \sqrt{(a + h_a\cot\alpha)^2 + h_a^2} = \sqrt{a^2 + b^2 - 2ab\cos\beta}$$
$$d_2 = \sqrt{(a - h_a\cot\alpha)^2 + h_a^2} = \sqrt{a^2 + b^2 - 2ab\cos\alpha}$$

$$d_1^2 + d_2^2 = 2(a^2 + b^2)$$

29. Rhombus
A rhombus is a closed planar parallelogram with all sides equal in length.

a) Area:

$$A = a^2\sin\alpha = a^2\sin\beta = ah_a = \frac{d_1 d_2}{2}$$

GEOMETRY
Areas

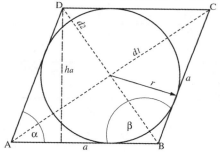

b) Diagonals:

$$d_1 = 2a\cos\frac{\alpha}{2}; \; d_2 = 2a\sin\frac{\alpha}{2}$$

$$d_1^2 + d_2^2 = 4a^2$$

c) Radius of inscribed circle:

$$r = \frac{d_1 d_2}{2\sqrt{d_1^2 + d_2^2}}.$$

d) Altitude:

$$h_a = a\sin\alpha$$

30. Trapezoid (American definition)

A trapezoid is a quadrilateral with one and only one pair of parallel sides.

a) Perimeter:

$$P = a + b + c + d$$

GEOMETRY
Areas

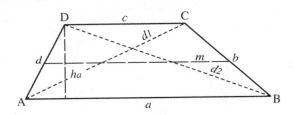

b) Area:

$$A = \frac{a+c}{2} h_a = m h_a$$

$$m = \frac{a+c}{2} \qquad (\,a \neq c,\ c \parallel a\,)$$

c) Altitude:

$$h_a^2 = \frac{k \cdot l}{4(a-c)^2}$$

where

$$k = (a+d-c+b) \cdot (d+c+b-a)$$
$$l = (a-d-c+b) \cdot (a+d-c-b)$$

31. Kite

A kite is a closed planar quadrilateral whose two pairs of distinct adjacent sides are equal in length. One diagonal bisects the other. Diagonals intersect at right angles.

a) Perimeter:

$$P = a+b+c+d$$

GEOMETRY
Areas

where

a, b, c, d = sides of kite $(a = b, \quad c = d)$

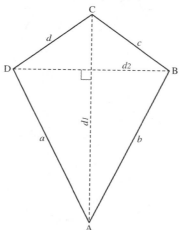

b) Area:

$$A = \frac{d_1 d_2}{2}$$

where

d_1, d_2 = diagonals of kite $(d_1 \perp d_2)$

32. Regular Polygon

A polygon is a closed plane figure with n sides. If all sides and angles are equal, the polygon is called regular.

a) Perimeter:

$$P = n \cdot a$$

GEOMETRY
Areas

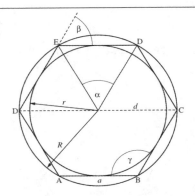

b) Area:

$$A = \frac{n}{2} R^2 \sin \alpha$$

c) Radius of circumscribed circle:

$$R = \frac{a}{2 \sin \dfrac{\alpha}{2}}$$

d) Radius of inscribed circle:

$$r = \frac{a}{2 \tan \dfrac{\alpha}{2}}$$

e) Central angles:

$$\alpha = \frac{360°}{n}$$

f) Internal angles:

$$\gamma = 180° - \beta = \frac{n-2}{n} \cdot 180°$$

GEOMETRY
Areas

g) External angles:

$$\beta = \alpha$$

h) Number of diagonals:

$$N = \frac{1}{2}n(n-3)$$

33. Circle

All points on the circumference of a circle are equidistant from its center.

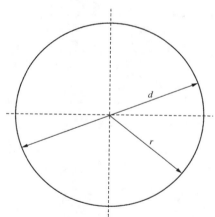

a) Perimeter:

$$P = 2\pi r = \pi d$$

b) Area:

$$A = \frac{\pi}{4}d^2 = \pi r^2$$

GEOMETRY
Areas

34. Sector of a Circle

A sector of a circle of radius r is the interior portion of the circle determined by a central angle α

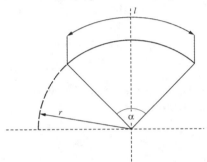

a) Angle:

$$\widehat{\alpha} = \frac{\pi}{180°} \alpha° \ [\text{rad}]$$

b) Length of arc:

$$l = \frac{\pi}{180°} r\alpha$$

c) Area:

$$A = \frac{\pi}{360°} r^2 \alpha°$$

35. Segment of a Circle

A segment is a portion of a circle whose upper boundary is a circular arc l and lower boundary is a secant s.

GEOMETRY
Areas

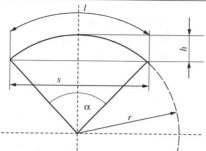

a) Angle:

$$\hat{\alpha} = \frac{\pi}{180^\circ} \alpha^\circ \ [\text{rad}]$$

b) Radius:

$$r = \frac{h}{2} + \frac{s}{8h}$$

c) Secant:

$$s = 2r\sin\frac{\alpha}{2} = 2\sqrt{h(2r-h)}$$

d) Area:

$$A = \frac{r^2}{2}\left(\hat{\alpha} - \sin\alpha^\circ\right) \approx \frac{h}{6s}\left(3h^2 + 4s^2\right)$$

e) Height:

$$h = r\left(1 - \cos\frac{\alpha^\circ}{2}\right) = \frac{s}{2}\tan\frac{\alpha^\circ}{4}$$

GEOMETRY
Areas

---

36. Annulus (Circular Ring)

The annulus is the plane area between two concentric circles, making a flat ring.

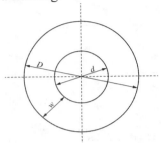

Area:

$$A = \frac{\pi}{4}\left(D^2 - d^2\right) = \pi(d + w)w$$

37. Ellipse

An ellipse is the locus of a point that moves in such a way that the sum of its distance from two fixed points (the foci) is constant.

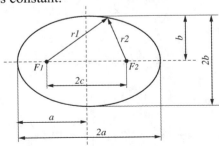

a) Radii:

$$r_1^2 + r_2^2 = 2a$$

b) Area:

$$A = \pi ab$$

c) Perimeter:

$$P \approx 2\pi \sqrt{\frac{1}{2}\left(a^2 + b^2\right)} = \pi(a+b)k$$

where

$$k = 1 + \frac{1}{4}m^2 + \frac{1}{64}m^4 + \frac{1}{256}\,m^6 + \dots \,, \quad m = \frac{a-b}{a+b}$$

38. Cube

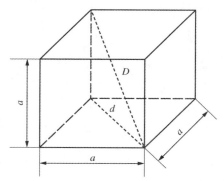

A cube is a regular hexahedron. It is composed of six square planes that meet each other at right angles; it has 12 edges.

a) Area:

$$A = 6a^2$$

b) Volume:
$$V = a^3$$
c) Diagonal of cube:
$$D = a\sqrt{3}$$
d) Diagonal of square face:
$$d = a\sqrt{2}$$

39. Cuboid

A cuboid is a solid body composed of three pairs of rectangular planes placed opposite each other and joined at right angles.

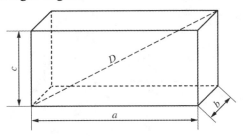

a) Area:
$$A = 2(ab + ac + bc)$$

b) Volume:
$$V = abc$$

c) Diagonal:
$$D = \sqrt{a^2 + b^2 + c^2}$$

GEOMETRY
Solid Bodies

40. Right Prism

A right prism is a solid body in which the bases (top and bottom) are identical polygons aligned such that the vertical face connecting their sides are rectangles at right angles to the bases.

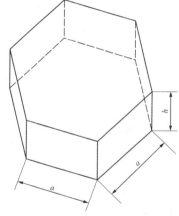

a) Volume: $V = A_b h$

b) Area: $A = 2A_b + A_l$

where

    A_b = area of base

    A_l = lateral area

41. Pyramid

A pyramid is a solid body whose base is a polygon and whose other faces are all triangles meeting at the apex. A right pyramid has its apex directly above the center of the base.

GEOMETRY
Solid Bodies

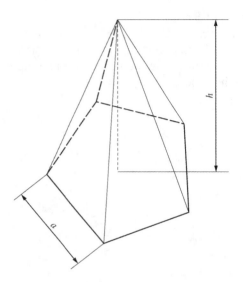

a) Area:

$$A = A_b + A_l$$

where

A_b = area of base

A_l = the lateral area

42. Frustum of Pyramid

Slicing the top off a pyramid leaves a frustum of a pyramid. It is determined by the plane of the base and a plane parallel to the base.

GEOMETRY
Solid Bodies

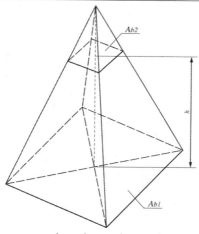

a) Area: $\qquad A = A_{b1} + A_{b2} + A_l$

where

$\qquad A_{b1}, A_{b2}$ = area of bases

$\qquad A_l$ = the lateral area

$\qquad h$ = height of pyramid

b) Volume:

$$V = \frac{h}{3}(A_{b1} + A_{b2} + \sqrt{A_{b1} \cdot A_{b2}})$$

43. Cone

A cone is a solid of the form described by the revolution of a right-angled triangle about one of the sides adjacent to the right angle, also called a right cone.

GEOMETRY
Solid Bodies

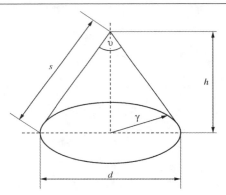

a) Volume:

$$V = \frac{A_b h}{3} = \frac{\pi}{3} \, r^2 h$$

b) Area:

$$A = A_b + A_l = \pi \cdot r^2 + \pi \cdot r \cdot s = \pi \cdot r (r + s)$$

where:

A_b = area of base

A_l = the lateral area.

s = slant height

c) Lateral area:

$$A_l = \pi \cdot r \cdot s$$

d) Area of base:

$$A_b = \frac{d^2 \pi}{4} = \pi r^2$$

e) Slant height:

$$s = \sqrt{r^2 + h^2}$$

f) Vertex angle:

$$\vartheta = 2\tan^{-1}\left(\frac{r}{h}\right)$$

g) Height:

$$h = \sqrt{s^2 - r^2}$$

44. Frustum of Cone

Slicing the top off a cone leaves a frustum of a cone. The plane of the base and a plane parallel to the base determine it.

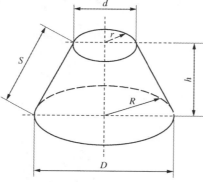

a) Area:

$$A = A_{b1} + A_{b2} + A_l$$
$$A = \pi\left[R^2 + r^2 + (R+r)s\right]$$

GEOMETRY
Solid Bodies

---

b) Area of bases:
$$A_{b1} = \pi R^2, \quad A_{b2} = \pi \cdot r^2$$

c) Lateral area:
$$A_l = \pi \cdot s(R + r)$$

d) Slant height:
$$s = \sqrt{h^2 + (R - r)^2}$$

e) Volume:
$$V = \frac{1}{3}\pi \cdot h \cdot \left(R^2 + r^2 + R \cdot r\right)$$

45. Cylinder

A cylinder is a solid body with a circular base and straight sides.

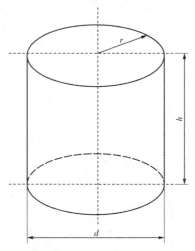

GEOMETRY
Solid Bodies

a) Area:

$$A = 2A_b + A_l$$

$$A = 2\pi \cdot r^2 + 2\pi \cdot r \cdot h$$

$$A = 2\pi \cdot r(r + h)$$

b) Area of base:

$$A_b = \frac{\pi \cdot d^2}{4} = \pi \cdot r^2$$

c) Lateral area:

$$A_l = 2\pi \cdot r \cdot h$$

d) Volume:

$$V = \frac{\pi}{4} d^2 h = \pi r^2 h$$

46. Hollow Cylinder

A hollow cylinder is a solid with circular ring bases and straight sides.

a) Volume:

$$V = A_b \cdot h = \frac{\pi}{4} h \cdot \left(D^2 - d^2\right)$$

where

A_b = annulus area

h = height of cylinder

D, d = outside and inside diameters of the hollow cylinder

GEOMETRY
Solid Bodies

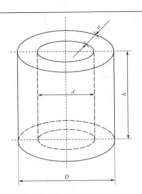

47. Sliced Cylinder

A sliced cylinder is a portion of a circular cylinder cut off by a sloped plane.

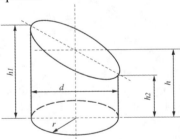

a) Area: $A = A_{b1} + A_{b2} + A_l$

$$A = \pi \cdot r \left[h_1 + h_2 + r + \sqrt{r^2 + \frac{(h_1 - h_2)^2}{4}} \right]$$

b) Lateral area: $A_l = \pi \cdot d \cdot h$

GEOMETRY
Solid Bodies

c) Volume: $\qquad V = \dfrac{\pi}{4} d^2 h$

48. Sphere

A sphere is defined as a three-dimensional figure with all of its points at the same distance r from its center.

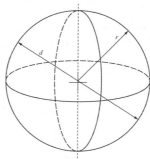

a) Area: $\qquad A = 4\pi \cdot r^2$

b) Volume: $\qquad V = \dfrac{4\pi \cdot r^3}{3}$

49. Spherical Cap

A spherical cap is a portion of a sphere cut off by a plane.

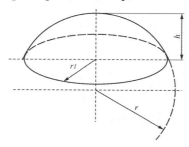

GEOMETRY
Solid Bodies

---

a) Area: $A = 2\pi r h = \pi\left(r_1^{\,2} + h^2\right)$

b) Volume: $V = \dfrac{1}{3}\pi h^2\left(3r - h\right)$

50. Sector of a Sphere

A sector of a sphere is the part of a sphere generated by a right circular cone that has its vertex at the center of the sphere.

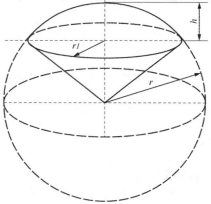

a) Area: $A = \pi \cdot r\left(2h + r_1\right)$

b) Volume: $V = \dfrac{2}{3}\pi \cdot r^2 \cdot h$

where

    r = radius of the sphere

GEOMETRY
Solid Bodies

---

r_1= radius of the base of the cone

51. Zone of a Sphere

The zone of a sphere is a portion cut off by two parallel planes.

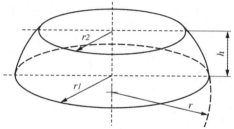

a) Area:

$$A = A_{b1} + A_{b2} + A_l$$

$$A = \pi r_1^2 + \pi r_2^2 + 2\pi \cdot r \cdot h$$

$$A = \pi\left(2rh + r_1^2 + r_2^2\right)$$

where

A_{b1}, A_{b2} = areas of bases

A_l = the lateral area of the zone

r_1, r_2 = radii of bases

b) Volume:

$$V = \frac{\pi}{6} h\left(3r_1^2 + 3r_2^2 + h^2\right)$$

GEOMETRY
Solid Bodies

---

52. Torus

A torus is the surface of a three-dimensional figure obtained by rotating a circle about an axis coplanar with the circle and at a fixed distance from the origin.

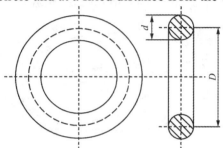

a) Area:

$$A = \pi^2 Dd$$

b) Volume:

$$V = \frac{\pi^2}{4} Dd^2$$

53. Ellipsoid

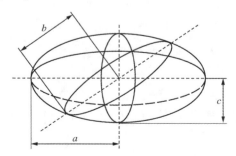

GEOMETRY
Solid Bodies

An ellipsoid is a three-dimensional figure, all planar cross-sections of which are ellipses. It is determined by three semi-axes: a, b, and c (where in general $a \neq b \neq c$). If two of those are equal, the ellipsoid is a spheroid; if all three are equal, it is a sphere.

a) Volume:

$$V = \frac{4}{3}\pi \cdot abc$$

where

a, b, c = semi-axes of ellipsoid

54. Barrel

A barrel is a solid that bulges out in the middle and has identical circular ends.

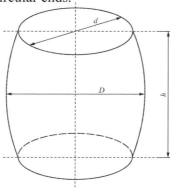

a) Volume: $\quad V = \frac{\pi}{12}h\left(2D^2 + d^2\right)$

TRIGONOMETRY

Trigonometry is the branch of mathematics concerned with solving problems that involve triangles, circles, oscillations, and waves using trigonometric ratios, which are seen as properties of triangles rather than of angles. It is absolutely crucial to much of geometry and physics.

This section contains:

- Fundamentals of Trigonometry
- Trigonometric Equations
- Graphs of the Trigonometric Functions

TRIGONOMETRY
Fundamentals of Trigonometry

---

1. Circular and Angular Measures

An angle is formed by two intersecting half-lines or by rotating a half-line from position OP to a terminal position OR. If the rotation is clockwise, the angle is deemed negative, and if counterclockwise the angle is deemed positive.

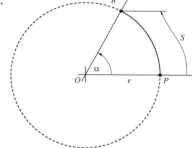

a) Circular measure

The circular measure is the ratio of the arc $PR = s$ to the radius r:

$$\widehat{\alpha} = \frac{s}{r} = 1 \ (\text{rad})$$

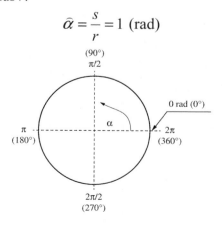

TRIGONOMETRY
Fundamentals of Trigonometry

---

b) Angular measure

The angular degree, symbolized by °, is a unit of plane angular measure. There are 360 angular degrees in a complete circle. Each degree is divided into 60 minutes and each minute is divided into 60 seconds.

c) Relation between circular and angular measures:

degrees	0°	30°	60°	90°	180°	270°	360°
radians	0	$\dfrac{\pi}{6}$	$\dfrac{\pi}{3}$	$\dfrac{\pi}{2}$	π	$\dfrac{3\pi}{2}$	2π
	0	0.52	1.05	1.57	3.14	4.71	6.28

1 radian = 57.2958 degrees

2. Trigonometric Circle

A circle centered at the origin O with radius 1 is called a trigonometric circle or unit circle.

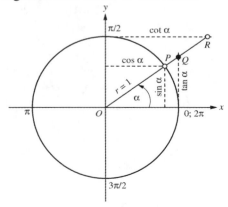

TRIGONOMETRY
Fundamentals of Trigonometry

The x-coordinate of point P is called the cosine of α.

The y-coordinate of point P is called the sine of α.

The y-coordinate of point Q is called the tangent of α.

The x-coordinate of point R is called the cotangent of α.

3. Basic Formulas

$$\sin\alpha \sin\beta = \frac{1}{2}\left[\cos(\alpha - \beta)\ \cos(\alpha + \beta)\right]$$

$$\cos\alpha \cos\beta = \frac{1}{2}\left[\cos(\alpha - \beta) + \cos(\alpha + \beta)\right]$$

$$\sin\alpha \cos\beta = \frac{1}{2}\left[\sin(\alpha - \beta) + \sin(\alpha + \beta)\right]$$

$$\sin^2 \alpha = \frac{1}{2}\left(1 - \cos 2\alpha\right)$$

$$\cos^2 \alpha = \frac{1}{2}\left(1 + \cos 2\alpha\right)$$

$$\sin^3 \alpha = \frac{1}{4}\left(3\sin\alpha - \sin 3\alpha\right)$$

$$\cos^3 \alpha = \frac{1}{4}\left(\cos 3\alpha + 3\cos\alpha\right)$$

TRIGONOMETRY
Fundamentals of Trigonometry

$$\sin^2\alpha + \cos^2\alpha = 1$$

$$\sin\alpha = \frac{1}{\csc\alpha}$$

$$\cos\alpha = \frac{1}{\sec\alpha}$$

$$\csc\alpha = \frac{1}{\sin\alpha}$$

$$\sec\alpha = \frac{1}{\cos\alpha}$$

$$\tan\alpha = \frac{\sin\alpha}{\cos\alpha} = \frac{1}{\text{ctn}\,\alpha}$$

$$\text{ctn}\,\alpha = \frac{\cos\alpha}{\sin\alpha} = \frac{1}{\tan\alpha}$$

$$\cos(-\alpha) = \cos\alpha$$

$$\sin(-\alpha) = -\sin\alpha$$

$$\sin\alpha = \sin(\alpha + 2\pi)$$

$$\cos\alpha = \cos(\alpha + 2\pi)$$

$$1 + \tan^2\alpha = \sec^2\alpha$$

$$1 + \text{ctn}^2\alpha = \csc^2\alpha = \frac{1}{\sin^2\alpha}$$

TRIGONOMETRY
Fundamentals of Trigonometry

If α and α' are supplementary angles $\left(\alpha + \alpha' = \pi\right)$, then

$$\sin \alpha = \sin(\alpha')$$
$$\cos \alpha = -\cos(\alpha')$$
$$\tan \alpha = -\tan(\alpha')$$
$$\text{ctn}\,\alpha = -\text{ctn}(\alpha')$$

If α and α' are complementary angles $\left(\alpha + \alpha' = \dfrac{\pi}{2}\right)$, then

$$\sin \alpha = \cos(\alpha')$$
$$\cos \alpha = \sin(\alpha')$$
$$\tan \alpha = \text{ctn}(\alpha')$$
$$\text{ctn}\,\alpha = \tan(\alpha')$$

If α and α' are opposite values $\left(\alpha + \alpha' = 0\right)$, then

$$\sin \alpha = -\sin(\alpha')$$
$$\cos \alpha = \cos(\alpha')$$
$$\tan \alpha = -\tan(\alpha')$$
$$\text{ctn}\,\alpha = -\text{ctn}(\alpha')$$

TRIGONOMETRY
Fundamentals of Trigonometry

---

4. Trigonometric Ratios for Right Angled Triangles

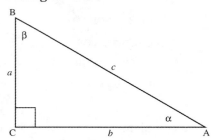

There are six ratios, defined as follows: three major and three *minor*.

Major:

$$\sin \alpha = \frac{\text{opp}}{\text{hyp}} = \frac{a}{c}$$

$$\cos \alpha = \frac{\text{adj}}{\text{hyp}} = \frac{b}{c}$$

$$\tan \alpha = \frac{\text{opp}}{\text{adj}} = \frac{a}{b}$$

Minor:

$$\operatorname{ctn} \alpha = \frac{\text{adj}}{\text{opp}} = \frac{b}{a}$$

$$\sec \alpha = \frac{\text{hyp}}{\text{adj}} = \frac{c}{b}$$

$$\operatorname{cosec} \alpha = \frac{\text{hyp}}{\text{opp}} = \frac{c}{a}$$

5. Sum and Difference of Functions of Angles

$$\sin \alpha + \sin \beta = 2 \sin \frac{\alpha + \beta}{2} \cos \frac{\alpha - \beta}{2}$$

$$\sin \alpha - \sin \beta = 2 \cos \frac{\alpha + \beta}{2} \sin \frac{\alpha - \beta}{2}$$

$$\cos \alpha - \cos \beta = -2 \sin \frac{\alpha + \beta}{2} \sin \frac{\alpha - \beta}{2}$$

TRIGONOMETRY
Fundamentals of Trigonometry

$$\cos\alpha + \cos\beta = 2\cos\frac{\alpha+\beta}{2}\cos\frac{\alpha-\beta}{2}$$

$$\tan\alpha \pm \tan\beta = \frac{\sin(\alpha\pm\beta)}{\cos\alpha\ \cos\beta}$$

$$\text{ctn}\alpha \pm \text{ctn}\beta = \frac{\sin(\alpha\pm\beta)}{\sin\alpha\ \sin\beta}$$

6. Sum and Difference of Angles

$$\sin(\alpha\pm\beta) = \sin\alpha\cos\beta \pm \cos\alpha\sin\beta$$

$$\cos(\alpha\pm\beta) = \cos\alpha\cos\beta \mp \sin\alpha\sin\beta$$

$$\tan(\alpha\pm\beta) = \frac{\tan\alpha\pm\tan\beta}{1\mp\tan\alpha\tan\beta}$$

$$\text{ctn}(\alpha\pm\beta) = \frac{\text{ctn}\alpha\,\text{ctn}\beta\mp1}{\pm\,\text{ctn}\alpha+\text{ctn}\beta}$$

7. Double Angle Formulas

$$\sin2\alpha = 2\sin\alpha\cos\alpha$$

$$\cos2\alpha = \cos^2\alpha - \sin^2\alpha$$

$$\tan2\alpha = \frac{2\tan\alpha}{1-\tan^2\alpha}$$

$$\text{ctn}2\alpha = \frac{\text{ctn}^2\alpha-1}{2\text{ctn}\alpha}$$

TRIGONOMETRY
Fundamentals of Trigonometry

8. Half Angle Formulas

$$\sin\frac{\alpha}{2} = \sqrt{\frac{1-\cos\alpha}{2}}$$

$$\cos\frac{\alpha}{2} = \sqrt{\frac{1+\cos\alpha}{2}}$$

$$\tan\frac{\alpha}{2} = \sqrt{\frac{1-\cos\alpha}{1+\cos\alpha}}$$

$$\operatorname{ctn}\frac{\alpha}{2} = \sqrt{\frac{1+\cos\alpha}{1-\cos\alpha}}$$

9. Functions of Important Angles

α	(°)	0°	30°	60°	90°	120°	180°
	rad	0	$\dfrac{\pi}{6}$	$\dfrac{\pi}{3}$	$\dfrac{\pi}{2}$	$\dfrac{2\pi}{3}$	π
$\sin\alpha$		0	$\dfrac{1}{2}$	$\dfrac{\sqrt{3}}{2}$	1	$\dfrac{\sqrt{3}}{2}$	0
$\cos\alpha$		1	$\dfrac{\sqrt{3}}{2}$	$\dfrac{1}{2}$	0	$-\dfrac{1}{2}$	-1

Continued from # 9

$\tan\alpha$	0	$\dfrac{\sqrt{3}}{3}$	$\sqrt{3}$	$\pm\infty$	$-\sqrt{3}$	0
$\operatorname{ctn}\alpha$	$\pm\infty$	$\sqrt{3}$	$\dfrac{\sqrt{3}}{3}$	0	$-\dfrac{\sqrt{3}}{3}$	$\pm\infty$

10. Solving Trigonometric Equations

Some equations that involve trigonometric functions of an unknown may be readily solved by using simple algebraic ideas, while others may be impossible to solve exactly but only approximately. Here are some methods for solving trigonometric equations:

a) Find the solution to the equation, and reduce to a base equation.

Example:

$$\tan\left(x - \frac{\pi}{2}\right) = \tan 2x \qquad (1)$$

Solution:

Reduce equation (1) to the base equation

$$\left(x - \frac{\pi}{2}\right) = 2x + k\pi$$

$$-x = \frac{\pi}{2} + k\pi$$

TRIGONOMETRY
Trigonometric Equations

It follows that

$$x = -\frac{\pi}{2} + k\pi$$

where k is any integer.

   b) Find the solution to the equation using
      factorization.

Example:
$$2\cos^2 x - 5\cos x + 2 = 0 \qquad (1)$$

Solution:
After factorization, the equation (1) has the form

$$(2\cos x - 1)(\cos x - 2) = 0$$

The roots of the equation are:
$$2\cos x - 1 = 0$$
$$2\cos x = 1$$
$$\cos x = \frac{1}{2}, \text{ and}$$

$$\cos x - 2 = 0$$
$$\cos x = 2$$

Remember that the range for $\cos x$ is

$$\left\{y: -1 \leq y \leq 1, \ y \text{ is real}\right\}$$

That is, y is between (-1) and 1, inclusive, so $\cos x \neq 2$.
Hence, the root that satisfies the equation (1) is

$$x = \frac{\pi}{3} + k\pi, \ k \in \text{ integer}$$

c) Find the solution to the equation using an additional unknown.

Example:
$$2 \sin^2(2x) + \sin(2x) - 1 = 0$$

Solution:

Let $u = \sin(2x)$
$$2u^2 + u - 1 = 0$$
$$u_{1,2} = -\frac{b \pm \sqrt{b^2 - 4ac}}{2a}$$
$$u_1 = \frac{1}{2}, \ u_2 = -1$$

Substitute:
$$\sin(2x) = \frac{1}{2}, \text{ or } \sin(2x) = -1$$

TRIGONOMETRY
Trigonometric Equations

$$\sin(2x) = \sin\frac{\pi}{6} \quad \text{or} \quad \sin(2x) = \sin\left(-\frac{3\pi}{2}\right)$$

$$2x = \frac{\pi}{6} + 2k\pi \quad \text{or} \quad 2x = \pi - \frac{\pi}{6} + 2k\pi$$

$$\text{or } 2x = \frac{3\pi}{2} + 2k\pi$$

$$x = \begin{cases} \dfrac{\pi}{12} + k\pi, \quad \dfrac{5\pi}{12} \, k\pi, \quad \text{or} \\ \dfrac{3\pi}{4} + k\pi, \quad k \in \text{ integer} \end{cases}$$

11. Verifying Trigonometric Identities

Example:

Verify the identity $\dfrac{1 + \tan x}{1 + \text{ctn} x} = \dfrac{\sin x}{\cos x}$

Solution:

Identities used:

$$\tan x = \frac{\sin x}{\cos x}; \quad \text{ctn} x = \frac{\cos x}{\sin x}$$

Divide the principal denominator into the principal numerator of the left term.

$$\frac{\cos x + \sin x}{\cos x} \cdot \frac{\sin x}{\sin x + \cos x} = \frac{\sin x}{\cos x}$$

TRIGONOMETRY
Trigonometric Equations

Reduce the left term by the factor $\sin x + \cos x$

$$\frac{\sin x}{\cos x} = \frac{\sin x}{\cos x}$$

Hence, the identities are correct.

Simplify the principal numerator and principal denominator of the left term.

$$\frac{1 + \dfrac{\sin x}{\cos x}}{1 + \dfrac{\cos x}{\sin x}} = \frac{\sin x}{\cos x}$$

Divide the principal denominator into the principal numerator of the left term.

$$\frac{\cos x + \sin x}{\cos x} \cdot \frac{\sin x}{\sin x + \cos x} = \frac{\sin x}{\cos x}$$

Reduce the left term by the factor $\sin x + \cos x$

$$\frac{\sin x}{\cos x} = \frac{\sin x}{\cos x}$$

Hence, the identity is correct.

TRIGONOMETRY
Graphs of the Trigonometric Functions

---

12. Graphs of the Sine and Cosine Functions

$y = \sin x$
$\quad$ for $\quad -\dfrac{\pi}{2} \leq x \leq 2\pi$
$y = \cos x$

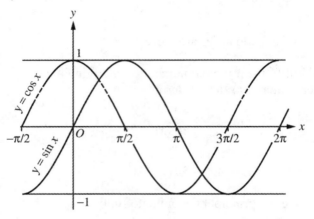

Domain: all real numbers.
Range: $-1 \leq y \leq 1$
Period: 2π

13. Graphs of the Tangent and Cotangent Functions

$y = \tan x$
$\qquad$ for $\quad -\pi < x < \pi$
$y = \operatorname{ctn} x$

TRIGONOMETRY
Graphs of the Trigonometric Functions

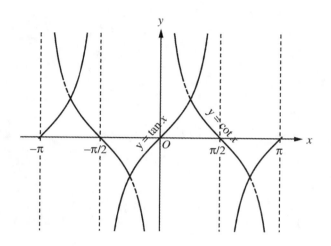

Domain: all real numbers except $\frac{\pi}{2} + k\pi$ for the tangent function, and $k\pi$, for the cotangent function (asymptotes occur here).

Range: all real numbers

Period: π

ANALYTICAL GEOMETRY

Analytic geometry, also called coordinate geometry, is the study of geometry using the principles of algebra. The Cartesian coordinate system is usually used to manipulate equations for planes, lines, curves, and circles, often in two but sometimes in three dimensions of measurement.

This section contains the most frequently used formulas, rules, and definitions relating to the following:

- Points and Lines
- Circles and Ellipses
- Parabolas
- Hyperbolas
- Polar Coordinates
- Introduction to Solid Analytical Geometry
- Planes
- Straight Lines in Space
- Surfaces

ANALYTICAL GEOMETRY
Points and Lines

1. Distance between Two Points

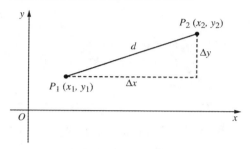

The distance between two points $P_1(x_1, y_1)$ and $P_2(x_2, y_2)$ is defined by the formula

$$d = \sqrt{(x_2 - x_1)^2 + (y_2 - y_1)^2}$$

where

$$\Delta x = x_2 - x_1$$
$$\Delta y = y_2 - y_1$$

2. Point of Division

The point of division is the point $P(x, y)$ which divides a line segment $P_1(x_1, y_1)$, $P_2(x_2, y_2)$ in a given ratio,

$$\lambda = \frac{P_1 P}{P P_2}$$

Point P has the coordinates

$$x = \frac{x_1 + \lambda x_2}{1 + \lambda}, \quad y = \frac{y_1 + \lambda y_2}{1 + \lambda}$$

ANALYTICAL GEOMETRY
Points and Lines

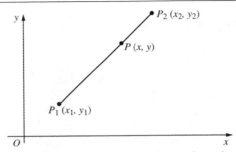

If $P(x, y)$ is the midpoint of line $P_1(x_1, y_1), P_2(x_2, y_2)$, $\lambda = 1$, then point P has the coordinates

$$x = \frac{x_1 + x_2}{2}, \quad y = \frac{y_1 + y_2}{2}$$

3. Inclination and Slope of a Line

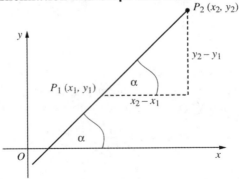

a) Inclination

The inclination of a line not parallel to the x-axis is defined as the smallest positive angle measured from the

ANALYTICAL GEOMETRY
Points and Lines

positive x-axis in a counterclockwise direction to the line. If the line is parallel to the *x*-axis, its inclination is defined as zero.

b) Slope

The slope of a line passing through two points $P_1(x_1, y_1)$ and $P_2(x_2, y_2)$ is

$$m = \tan \alpha = \frac{y_2 - y_1}{x_2 - x_1} = \frac{\Delta y}{\Delta x}$$

4. Parallel and Perpendicular Lines

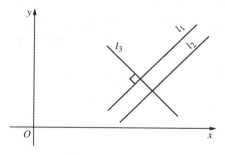

If line l_1 is parallel to line l_2, then their slopes are equal:

$$m_1 = m_2$$

If line l_1 and l_3 are perpendicular, the slope of one of the lines is the negative reciprocal of the slope of the other line.

If m_1 is the slope of l_1 and m_3 is the slope of l_3, then

ANALYTICAL GEOMETRY
Points and Lines

$$m_1 = -\frac{1}{m_3}, \quad \text{or} \quad m_1 \cdot m_3 = -1$$

5. Angle Between Two Intersecting Lines

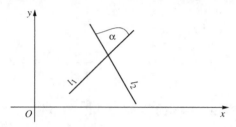

Angle α, measured in a positive direction counterclockwise from line l_1, whose slope is m_1, to line l_2, whose slope is m_2, is

$$\tan \alpha = \frac{m_2 - m_1}{1 + m_1 m_2}$$

6. Triangle

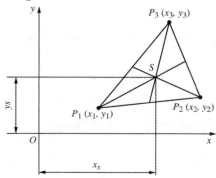

ANALYTICAL GEOMETRY
Points and Lines

The area of a triangle in terms of the vertices is

$$A = \frac{1}{2}\left(x_1 y_2 + x_2 y_3 + x_3 y_1 - x_3 y_2 - x_2 y_1 - x_1 y_3\right)$$

The coordinates of the centroid S (center of gravity) of the triangle are

$$x_s = \frac{x_1 + x_2 + x_3}{3}; \quad y_s = \frac{y_1 + y_2 + y_3}{3}$$

7. The Equation for a Straight Line through a Point

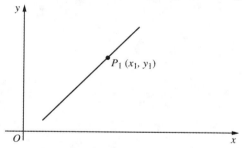

A straight line is completely determined if its gradient is known and a point $P(x_1, y_1)$ is given through which the line must pass:

$$y - y_1 = m(x - x_1)$$

8. Slope-Intercept Form

A straight line is determined if its slope (gradient) m and the y-intercept $(0, b)$ are known. Its equation is

ANALYTICAL GEOMETRY
Points and Lines

$$y = mx + b$$

9. Equation for a Straight Line through Two Points

The equation of a straight line through two defined points $P_1(x_1, y_1)$, and $P_2(x_2, y_2)$ is

$$\frac{y - y_1}{x - x_1} = \frac{y_1 - y_2}{x_1 - x_2}$$

10. General Form of an Equation of a Straight Line

$$Ax + By + C = 0$$

where

A, B and C are arbitrary constants.

For an equation in this form, the slope m and y-intercept b are

$$m = -\frac{A}{B}; \qquad b = -\frac{C}{B}$$

ANALYTICAL GEOMETRY
Points and Lines

---

11. Normal Equation of a Straight Line

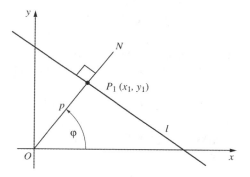

A straight line is determined if the length of the
perpendicular (p) from the origin to the line and the
angle (φ) which this perpendicular makes with the
x-axis are known.

The normal form of the equation of the straight line is

$$x \cos \varphi + y \sin \varphi - p = 0$$

The normal form of equation $Ax + By + C = 0$ is

$$\frac{A}{\pm \sqrt{A^2 + B^2}} x + \frac{B}{\pm \sqrt{A^2 + B^2}} y + \frac{C}{\pm \sqrt{A^2 + B^2}} = 0$$

where

$$\cos \varphi = \frac{A}{\pm \sqrt{A^2 + B^2}}; \quad \sin \varphi = \frac{B}{\pm \sqrt{A^2 + B^2}};$$

ANALYTICAL GEOMETRY
Points and Lines

$$-p = \frac{C}{\pm\sqrt{A^2 + B^2}}$$

12. Distance From a Line to a Point

The distance from a line l to a point $P_1\,(x_1, y_1)$ is the perpendicular distance d.

Since the coordinates of point $P_1\,(x_1, y_1)$ satisfy the equation for l_1,

$$x_1 \cos\varphi + y_1 \sin\varphi - (p + d) = 0$$

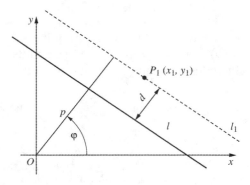

solving for d,

$$d = x_1 \cos\varphi + y_1 \sin\varphi - p,$$

or

$$d = \frac{|Ax_1 + By_1 + C|}{\sqrt{A^2 + B^2}}$$

ANALYTICAL GEOMETRY
Circles and Ellipses

---

13. Circles

A circle is represented by an equation of the second degree. A circle is completely defined if its center $M(p, q)$ and radius r are known:

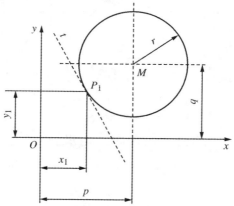

a) The equation of a circle:

$$(x - p)^2 + (y - q)^2 = r^2$$

If the center of a circle is at the origin, the equation becomes

$$x^2 + y^2 = r^2$$

The general equation of a circle is

$$x^2 + y^2 + Dx + Ey + F = 0, \text{ or}$$

ANALYTICAL GEOMETRY
Circles and Ellipses

$$\left(x+\frac{D}{2}\right)^2+\left(y+\frac{E}{2}\right)^2=\frac{D^2+E^2-4F}{4}$$

The center of the circle is at the point $M\left(-\frac{D}{2},-\frac{E}{2}\right)$,

and the radius of circle is

$$r=\frac{1}{2}\sqrt{D^2+E^2-4F}$$

If $D^2+E^2-4F>0$, the circle is real.

If $D^2+E^2-4F<0$, the circle is imaginary.

If $D^2+E^2-4F=0$, there is no circle (it is just the point M).

b) The tangent t at point $P_1(x_1, y_1)$:

$$y=\frac{r^2-(x-p)(x_1-p)}{y_1-q}+q$$

14. Ellipses

An ellipse is a curve in which the sum of the distances from any point on the curve to two fixed points is constant. The two fixed points are called foci (plural of focus).

ANALYTICAL GEOMETRY
Circles and Ellipses

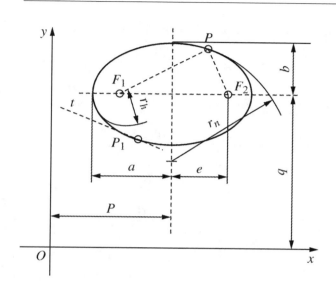

a) The equation of an ellipse:

$$\frac{(x-p)^2}{a^2} + \frac{(y-q)^2}{b^2} - 1 = 0$$

If the center is at the origin, the equation becomes

$$\frac{x^2}{a^2} + \frac{y^2}{b^2} = 1$$

In either case, the general form of the equation of the ellipse is

$$Ax^2 + By^2 + Dx + Ey + f = 0$$

b) Eccentricity:

$$e = \sqrt{a^2 - b^2} \qquad (a > b)$$

c) Vertex radii:

$$r_h = \frac{b^2}{a}, \quad r_n = \frac{a^2}{b}$$

d) Basic property:

$$\overline{F_1 P} + \overline{F_2 P} = 2a$$

where

$$F_1, F_2 = \text{focal points}$$

e) The equation of a tangent t at point $P_1(x_1, y_1)$:

$$y = -\frac{b^2}{a^2} \cdot \frac{(x_1 - p)(x - x_1)}{y_1 - q} + y_1$$

ANALYTICAL GEOMETRY
Parabolas

15. Parabolas

A parabola is the set of all points in a plane equidistant from a given line L (the conic section directrix) and a given point F not on the line (the focus). The focal parameter (i.e., the distance between the directrix and focus) is therefore given as p. The surface of revolution obtained by rotating a parabola about its axis of symmetry is called a paraboloid.

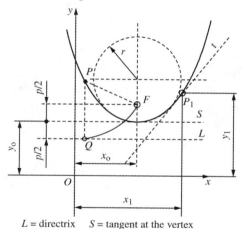

L = directrix S = tangent at the vertex

a) The equation of a parabola:

$$(x - x_0)^2 = 2p(y - y_0)$$

b) Basic equation:
$$y = ax^2 + bx + c$$

ANALYTICAL GEOMETRY
Hyperbolas

---

c) Vertex radius:
$$r = p$$

d) Basic property:
$$\overline{PF} = \overline{PQ}, \quad \frac{PF}{PQ} = 1 = e \text{ (eccentricity)}$$

e) Equation of a tangent at point $P_1(x_1, y_1)$:

$$y = \frac{2(y_1 - y_0)(x - x_1)}{x_1 - x_0} + y_1$$

16. Hyperbolas

A hyperbola is the set of all points $P(x, y)$ in the plane, the difference of whose distances from two fixed points F_1 and F_2 is some constant. The two fixed points are called the foci.

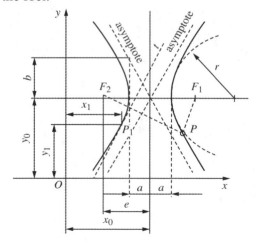

ANALYTICAL GEOMETRY
Hyperbolas

---

a) Equation of a hyperbola:

$$\frac{(x-x_0)^2}{a^2} - \frac{(y-y_0)^2}{b^2} - 1 = 0$$

If the point of intersection of the asymptotes is at the origin, the equation is

$$\frac{x^2}{a^2} - \frac{y^2}{b^2} - 1 = 0$$

b) Basic equation:

$$Ax^2 + By^2 + Cx + Dy + E = 0$$

c) Eccentricity:

$$e = \sqrt{a^2 + b^2}$$

d) The equation of asymptotes:

$$y = \pm \frac{b}{a} x$$

e) The equation of a tangent at point P_1 (x_1, y_1):

$$y = \frac{b^2}{a^2} \frac{(x_1 - x_0)(x - x_1)}{y_1 - y_0} + y_1$$

f) Vertex radius:

$$r = \frac{b^2}{a}$$

ANALYTICAL GEOMETRY
Polar Coordinates

---

17. Polar Coordinates

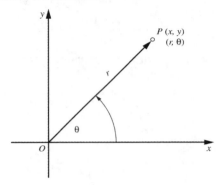

Let x and y be Cartesian axes in the plane and let P be a point in the plane other than the origin. The polar coordinates of point P are r (the radial coordinate) and θ (the angular coordinate, often called the polar angle), and they are defined in terms of Cartesian coordinates by

$$x = r\cos\theta$$

$$y = r\sin\theta$$

where

r = the radial distance ($r = OP > 0$)

θ = the counterclockwise angle from the x-axis

In terms of x and y they are

$$r = \sqrt{x^2 + y^2}$$

$$\theta = \tan^{-1}\left(\frac{y}{x}\right)$$

ANALYTICAL GEOMETRY
Introduction to Solid Analytical Geometry

18. Cartesian Coordinates

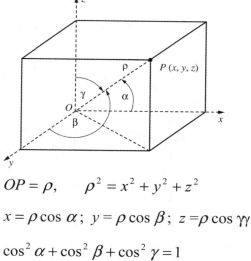

$$OP = \rho, \qquad \rho^2 = x^2 + y^2 + z^2$$

$$x = \rho \cos \alpha; \; y = \rho \cos \beta; \; z = \rho \cos \gamma\gamma$$

$$\cos^2 \alpha + \cos^2 \beta + \cos^2 \gamma = 1$$

$$\cos \alpha = \frac{x}{\rho}; \; \cos\beta = \frac{y}{\rho}; \; \cos\gamma = \frac{z}{\rho}$$

or

$$\cos\alpha = \frac{x}{\sqrt{x^2 + y^2 + z^2}}$$

$$\cos\beta = \frac{y}{\sqrt{x^2 + y^2 + z^2}}$$

$$\cos \gamma = \frac{z}{\sqrt{x^2 + y^2 + z^2}}$$

ANALYTICAL GEOMETRY
Introduction to Solid Analytical Geometry

---

19. Distance between Two Points

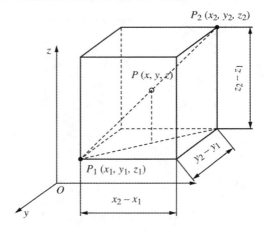

a) Distance between two points P_1 and P_2:

$$d = \sqrt{\left(x_2 - x_1\right)^2 + \left(y_2 - y_1\right)^2 + \left(z_2 - z_1\right)^2}$$

b) Point of division
If the point $P\,(x, y, z)$ divides the line $P_1(x_1, y_1, z_1)$ to

$P_2\,(x_2, y_2, z_2)$ in the ratio $\dfrac{P_1 P}{P P_2} = \dfrac{r}{1}$, then

$$x = \frac{x_1 + r x_2}{1 + r}, \quad y = \frac{y_1 + r y_2}{1 + r}, \quad z = \frac{z_1 + r z_2}{1 + r}$$

ANALYTICAL GEOMETRY
Introduction to Solid Analytical Geometry

c) Direction of a line
The direction cosines of P_1P_2 are

$$\cos \alpha = \frac{x_2 - x_1}{\sqrt{(x_2 - x_1)^2 + (y_2 - y_1)^2 + (z_2 - z_1)}}$$

$$\cos \beta = \frac{y_2 - y_1}{\sqrt{(x_2 - x_1)^2 + (y_2 - y_1)^2 + (z_2 - z_1)}}$$

$$\cos \gamma = \frac{z_2 - z_1}{\sqrt{(x_2 - x_1)^2 + (y_2 - y_1)^2 + (z_2 - z_1)}}$$

20. Angle between Two Lines
The angle between two lines that do not meet is defined as the angle between two intersecting lines, each of which is parallel to one of the given lines.

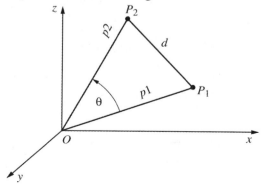

ANALYTICAL GEOMETRY
Introduction to Solid Analytical Geometry

If OP_1 and OP_2 are two lines through the origin parallel to the two given lines, and θ is the angle between the lines, from triangle OP_1P_2, by the law of cosines,

$$\cos\theta = \frac{\rho_1^2 + \rho_2^2 - d^2}{2\rho_1\rho_2}$$

where

$$\rho_1^2 = x_1^2 + y_1^2 + z_1^2$$
$$\rho_2^2 = x_2^2 + y_2^2 + z_2^2$$
$$d^2 = (x_2 - x_1)^2 + (y_2 - y_1)^2 + (z_2 - z_1)^2$$

Substituting and simplifying gives

$$\cos\theta = \frac{x_1 x_2 + y_1 y_2 + z_1 z_2}{\rho_1\rho_2}$$

But

$$\cos\alpha_1 = \frac{x_1}{\rho_1}, \ \cos\alpha_2 = \frac{x_2}{\rho_2}$$

$$\cos\beta_1 = \frac{y_1}{\rho_1}, \ \cos\beta_2 = \frac{y_2}{\rho_2}$$

$$\cos\gamma_1 = \frac{z_1}{\rho 1}, \ \cos\gamma_2 = \frac{z_2}{\rho_2}$$

Hence,

$$\cos\theta = \cos\alpha_1 \cos\alpha_2 + \cos\beta_1 \cos\beta_2 + \cos\gamma_1 \cos\gamma_2$$

21. Planes

Every plane can be represented by an equation of the first degree in one or more of the variables x, y, and z.

a) The equation of a plane:

$$Ax + By + Cz + D = 0 , \qquad (A, B, C) \neq 0$$

b) The equation of a system of planes passing through a point (x_0, y_0, z_0):

$$A(x - x_0) + B(y - y_0) + C(z - z_0) = 0$$

22. Line Perpendicular to a Plane

A line is perpendicular to a plane $Ax + By + Cz + D = 0$ if and only if the direction numbers a, b, c of the line are proportional to the coefficients of x, y, z in the equation of the plane. Hence:

$$\frac{a}{A} = \frac{b}{B} = \frac{c}{C} , \qquad (a, b, c, A, B, C) \neq 0$$

23. Parallel and Perpendicular Planes

a) Given two planes

$$A_1 x + B_1 y + C_1 z + D_1 = 0,$$
$$A_2 x + B_2 y + C_2 z + D_2 = 0,$$

ANALYTICAL GEOMETRY
Planes

---

the planes are parallel if and only if the coefficients of
$x, y, z,$ are proportional. Hence,

$$\frac{A_1}{A_2} = \frac{B_1}{B_2} = \frac{C_1}{C_2}$$

b) Two planes are perpendicular if
$$A_1 A_2 + B_1 B + C_1 C_2 = 0$$

24. Distance of a Point from a Plane
The distance between a point $P_1(x_1, y_1, z_1)$ and a plane
$Ax + By + Cz + D = 0$ is

$$d = \left| \frac{Ax_1 + By_1 + Cz_1 + D}{\sqrt{A^2 + B^2 + C^2}} \right|$$

25. Normal Form
The normal form of the equation of a plane is

$$x \cos \alpha + y \cos \beta + z \cos \gamma - p = 0$$

where

$p =$ the perpendicular distance from the origin to the plane

$a, \beta, \gamma =$ the direction angles of that perpendicular distance

ANALYTICAL GEOMETRY
Planes

The normal form of the equation of the plane

$$Ax + By + Cz + D = 0 \text{ is}$$

$$\frac{Ax + By + Cz + D}{\pm\sqrt{A^2 + B^2 + C^2}} = 0$$

The sign of the radical is taken opposite to that of D so that the normal distance p will be positive.

26. Intercept Form
The intercept form of the equation of a plane is

$$\frac{x}{a} + \frac{y}{b} + \frac{z}{c} = 1$$

where

$a, b, c =$ the x, y, z intercepts respectively.

27. Angle between Two Planes

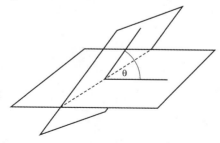

The angle between two planes

ANALYTICAL GEOMETRY
Straight Lines in Space

$$A_1 x + B_1 y + C_1 z + D_1 = 0,$$
$$A_2 x + B_2 y + C_2 z + D_2 = 0$$

is determined by

$$\cos\theta = \frac{A_1 A_2 + B_1 B_2 + C_1 C_2}{\sqrt{A_1^2 + B_1^2 + C_1^2} \sqrt{A_2^2 + B_2^2 + C_2^2}}$$

28. Straight Line in Space

The line of intersection of two planes

$$A_1 x + B_1 y + C_1 z + D_1 = 0,$$
$$A_2 x + B_2 y + C_2 z + D_2 = 0$$

is a straight line in space.

29. Parametric Form Equations of a Line

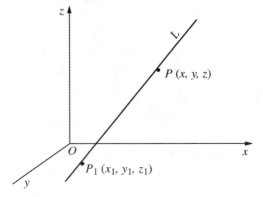

ANALYTICAL GEOMETRY
Straight Lines in Space

$$x = x_1 + \lambda \cos \alpha$$
$$y = y_1 + \lambda \cos \beta$$
$$z = z_1 + \lambda \cos \gamma$$

or

$$x = x_1 + a\lambda, \;\; y = y_1 + b\lambda, \;\; z = z_1 + c\lambda$$

where

α, β, γ = the direction angles of the line L

a, b, c = the direction numbers of the line L

λ = the variable length P_1P

30. Symmetric Form Equations of a Line

The equations of the line passing through point $P_1(x_1, y_1, z_1)$ have the form

$$\frac{x - x_1}{\cos \alpha} = \frac{y - y_1}{\cos \beta} = \frac{z - z_1}{\cos \gamma}$$

or

$$\frac{x - x_1}{a} = \frac{y - y_1}{b} = \frac{z - z_1}{c}$$

where

α, β, γ = the direction angles of the line

a, b, c = the direction numbers of the line

---

31. Two-Point Form Equations of a Line

The equations of the straight line through points $P1(x_1, y_1, z_1)$ and $P_2(x_2, y_2, z_2)$ are

$$\frac{x - x_1}{x_2 - x_1} = \frac{y - y_1}{y_2 - y_1} = \frac{z - z_1}{z_2 - z_1}$$

32. Relative Directions of a Line and Plane

A line whose direction numbers are a, b, c and the plane $Ax + By + Cz + D = 0$ are:

a) Parallel if and only if

$$Aa + Bb + Cc = 0$$

b) Perpendicular if and only if

$$\frac{A}{a} = \frac{B}{b} = \frac{C}{c}$$

33. Sphere

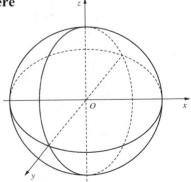

ANALYTICAL GEOMETRY
Surfaces

A sphere is a three-dimensional surface, all points of which are equidistant from a fixed point called the center. The equation of a sphere with center at $(0, 0, 0)$ and radius r is

$$x^2 + y^2 + z^2 = r^2$$

If the center of the sphere is at (h, k, j), the equation has the form

$$(x - h)^2 + (y - k)^2 + (z - j)^2 = r^2$$

34. Ellipsoid

An ellipsoid is a three-dimensional surface, all plane sections of which are ellipses or circles.

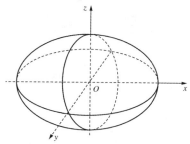

The equation of an ellipsoid with center at $(0, 0, 0)$ and semi-principal axes of unequal lengths a, b, and c is

$$\frac{x^2}{a^2} + \frac{y^2}{b^2} + \frac{z^2}{c^2} = 1$$

If $a \neq b$ but $b = c$, the ellipsoid is an ellipsoid of revolution. If the center of the ellipsoid is at (h, k, j) and its axes are parallel to the coordinate axes, the equation has the form

ANALYTICAL GEOMETRY
Surfaces

$$\frac{(x-h)^2}{a^2} + \frac{(y-k)^2}{b^2} + \frac{(z-j)^2}{c^2} = 1$$

If the center of the ellipsoid is at the origin, this equation becomes

$$\frac{x^2}{a^2} + \frac{y^2}{b^2} + \frac{z^2}{c^2} = 1$$

35. Hyperboloid

A hyperboloid is a quadric surface generated by rotating a hyperbola around its main axis.

a) Hyperboloid of one sheet:

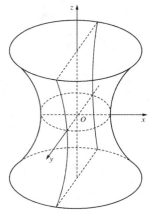

If the equation of the ellipsoid has the sign of one variable changed, as in

$$\frac{x^2}{a^2} + \frac{y^2}{b^2} - \frac{z^2}{c^2} = 1$$

ANALYTICAL GEOMETRY
Surfaces

---

the surface is called a hyperboloid of the sheet. If a = b, the surface is a hyperboloid of revolution of one sheet.

b) Hyperboloid of two sheets:

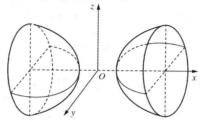

The equation of a hyperboloid of two sheets is

$$\frac{x^2}{a^2} - \frac{y^2}{b^2} - \frac{z^2}{c^2} = 1$$

36. Elliptic Paraboloid

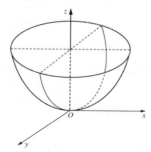

This is the locus of an equation of the form

$$\frac{x^2}{a^2} + \frac{y^2}{b^2} = 2cz$$

The section by a plane $z = k$ is an ellipse that increases in size as the cutting plane recedes from the xy-plane. If $c > 0$, the surface lies wholly above the xy-plane. If $c < 0$, the surface lies wholly below the xy-plane. If $a = b$, the surface is a surface of revolution.

37. Hyperbolic Paraboloid

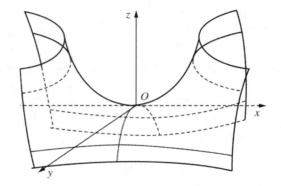

This is the locus of an equation of the form

$$\frac{x^2}{a^2} - \frac{y^2}{b^2} = 2cz \qquad (c > 0)$$

MATHEMATICS OF FINANCE

Financial mathematics is the application of mathematical methods to the solution of problems in finance. Many people are in the dark when it comes to applying math to practical problem solving. This section will show you how to do the math required to figure out a home mortgage, automobile loan, or the present value of an annuity, to compare investment alternatives, and much more.

This section contains formulas, definitions and some examples regarding:

- Simple Interest
- Compound Interest
- Annuities
- Amortization

1. Simple Interest

Interest is the fee paid for the use of someone else's money. Simple interest is interest paid only on the amount deposited and not on past interest accumulated. The formula for simple interest is

$$I = P \cdot r \cdot t$$

where

I = interest
P = principal
r = annual interest rate in percent
t = time in years

Example:
Find the simple interest for $1,500 at 8% for 2 years.
Solution:

P = $1,500, r = 8% = 0.08, and t = 2 years
$I = P \cdot r \cdot t$ = (1500)(0.08)(2) = 240 or $240

a) Future value

If P dollars are deposited at interest rate r for t years, the money earns interest. When this interest is added to the initial deposit, the total amount in the account is

$$A = P + I = P + Prt = P(1 + rt)$$

This amount is called the future value or maturity value.

MATHEMATICS OF FINANCE
Compound Interest

Example:
Find the maturity value of $10,000 at 8% for 6 months.

Solution:
$P = \$10,000$, $r = 8\% = 0.08$, $t = 6/12 = 0.5$ years.
The maturity value is
$A = P(1 + rt) = 10,000[1 + 0.08(0.5)] = 10400$, or $10,400

2. Compound Interest

Simple interest is normally used for loans or investment of a year or less. For longer periods, compound interest is used. The compound amount at the end of t years is given by the compound interest formula,

$$A = P(1 + i)^n$$

where

i = interest rate per compounding period ($i = \dfrac{r}{m}$)

n = number of conversion periods for t years
$(n = mt)$

A = compound amount at the end of n conversion period

P = principal

r = nominal interest rate per year

MATHEMATICS OF FINANCE
Compound Interest

---

m = number of conversion (compounding) periods per year
t = term (number of years)

Example:
Suppose \$15,000 is deposited at 8% and compounded annually for 5 years. Find the compound amount.

Solution:
$$P = \$15{,}000, \; r = 8\% = 0.08, \; m = 1, \; n = 5$$

$$A = P(1+i)^n = 15000\left[1+\left(\frac{0.08}{1}\right)\right]^5 = 15000 \cdot [1.08]^5$$

$$= 22039.92, \text{ or } \$22{,}039.92$$

a) Continuous compound interest
The compound amount A for a deposit of P at interest rate r per year compounded continuously for t years is given by

$$A = Pe^{rt}$$

where

P = principal
r = annual interest rate compounded continuously
t = time in years
A = compound amount at the end of t years.
e = 2.7182818

MATHEMATICS OF FINANCE
Compound Interest

---

b) Effective rate

The effective rate is the simple interest rate that would produce the same accumulated amount in one year as the nominal rate compounded m times a year. The formula for effective rate of interest is

$$r_{eff} = \left(1 + \frac{r}{m}\right)^m - 1$$

where

r_{eff} = effective rate of interest

r = nominal interest rate per year

m = number of conversion periods per year

Example:

Find the effective rate of interest corresponding to a nominal rate of 8% compounded quarterly.

Solution:

$r = 8\% = 0.08$, $m = 4$, then

$$r_{eff} = \left(1 + \frac{r}{m}\right)^m - 1 = \left(1 + \frac{0.08}{4}\right)^4 - 1 = 0.082432,$$

so the corresponding effective rate on this case is 8.243% per year.

c) Present value with compound interest
The principal P is often referred to as the present value, and the accumulated value A is called the future value since it is realized at a future date. The present value is given by

$$P = \frac{A}{(1+i)^n} = A(1+i)^{-n}$$

Example:
How much money should be deposited in a bank paying interest at the rate of 3% per year compounding monthly so that at the end of 5 years the accumulated amount will be $15,000?

Solution:
Here:
- nominal interest per year $r = 3\% = 0.03$,
- number of conversion periods per year $m = 12$,
- interest rate per compounding period
  $i = 0.03/12 = 0.0025$,
- number of conversion periods for t years
  $n = (5)(12) = 60$,
- accumulated amount $A = 15,000$

$$P = A(1+i)^{-n} = 15,000(1+0.0025)^{-60}$$
$$P = 12,913.04, \text{ or } \$12,913$$

3. Annuities

An annuity is a sequence of payments made at regular time intervals. This is the typical situation when finding the relationship between the amount of money loaned and the size of the payments.

a) Present value of annuity

The present value P of an annuity of n payments of R dollars each, paid at the end of each investment period into an account that earns interest at the rate of i per period, is

$$P = R\left[\frac{1 - (1+i)^{-n}}{i}\right]$$

where

P = present value of annuity

R = regular payment per month

n = number of conversion periods for t years

i = interest rate per conversion period

Example:

What size loan could Bob get if he can afford to pay \$1,000 per month for 30 years at 5% annual interest?

Solution:

Here: $R = 1,000$

$i = 0.05/12 = 0.004167$

$n = (12)(30) = 360$

MATHEMATICS OF FINANCE
Annuities

$$P = R\left[\frac{1-(1+i)^{-n}}{i}\right] = 1000\left[\frac{1-(1+0.05/12)^{-360}}{0.05/12}\right]$$

$P = 186281.62$ or $\$186,281.62$

Under these terms, Bob would end up paying a total of $360,000, so the total interest paid would be $360,000 − $186,281.62 = $173,713.38.

b) Future value of an annuity

The future value S of an annuity of n payments of R dollars each, paid at the end of each investment period into an account that earns interest at the rate of i per period, is

$$S = R\left[\frac{(1+i)^n - 1}{i}\right]$$

Example:

Let us consider the future value of $1,000 paid at the end of each month into an account paying 5% annual interest for 30 years. How much will accumulate?

Solution:

This is a future value calculation with $R = 1,000$, $n = 360$, and $i = 0.05/12 = 0.004167$. This account will accumulate as follows

MATHEMATICS OF FINANCE
Amortization

$$S = R\left[\frac{(1+i)^n - 1}{i}\right] = 1000\left[\frac{(1+0.05/12)^{360} - 1}{0.05/12}\right]$$

$$S = 832258.64, \text{ or } \$832,258.64$$

Note: This is much larger than the sum of the payments, since many of those payments are earning interest for many years.

4. Amortization of Loans

The periodic payment R_a on a loan of P dollars to be amortized over n periods with interest charged at the rate of i per period is

$$R_a = \frac{Pi}{1 - (1+i)^{-n}}$$

Example:

Bob borrowed $120,000 from a bank to buy a house. The bank charges interest at a rate of 5% per year. Bob has agreed to repay the loan in equal monthly installments over 30 years. How much should each payment be if the loan is to be amortized at the end of that time?

Solution:

This is a periodic payment calculation with $P = 120,000$, $i = 0.05/12 = 0.004167$, and

---

$n = (30)(12) = 360$

$$R_a = \frac{Pi}{1-(1+i)^{-n}} = \frac{(120000)(0.05/12)}{1-(1+0.05/12)^{-360}} = 644.19$$

or $644.19.

5. Sinking Fund Payment

The sinking fund calculation is used to calculate the periodic payments that will accumulate by a specific future date to a specified future value, so that investors can be certain that the funds will be available at maturity.

The periodic payment Ra needed to reach an accumulated amount S after n periods with an interest rate of i per period is

$$R_a = \frac{Si}{(1+i)^n - 1}$$

where

S = the future value
i = interest rate per period
n = number of conversion periods for t years

CALCULUS

Calculus is a branch of mathematics developed from algebra and geometry and built on two major complementary ideas.

One concept is *differential calculus*. It studies rates of change, such as how fast an airplane is going at any instant after take-off, the acceleration and speed of a free-falling body at a particular moment, etc.

The other key concept is *integral calculus*. It studies the accumulation of quantities, such as areas under a curve, linear distance traveled, or volume displaced.

Integral calculus is the mirror image of differential calculus.

This section contains:

- Limits
- Derivatives
- Integration
- Basic Integrals
- Applications of Integration

CALCULUS
Limits

1. Limits

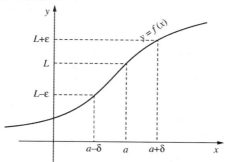

If the value of the function $y = f(x)$ gets arbitrarily close to L as x approaches the point a, then we say that the limit of the function as x approaches a is equal to L. This is written as

$$\lim_{x \to a} f(x) = L$$

2. Rules for Limits

Let u and v be functions such that

$$\lim_{x \to a} u(x) = A \quad \text{and} \quad \lim_{x \to a} v(x) = B$$

1) $\lim_{x \to a}[ku(x) \pm hv(x)] =$

   $k \lim_{x \to a} u(x) \pm h \lim_{x \to a} v(x) = kA \pm kB$

2) $\lim_{x \to a}[u(x) \cdot v(x)] = \left[\lim_{x \to a} u(x)\right] \cdot \left[\lim_{x \to a} v(x)\right] = A \cdot B$

---

3) $\lim\limits_{x \to a} \dfrac{u(x)}{v(x)} = \dfrac{\lim\limits_{x \to a} u(x)}{\lim\limits_{x \to a} v(x)} = \dfrac{A}{B}$ $(B \neq 0)$

4) $\lim\limits_{x \to a} [u(x)]^n = \left[\lim\limits_{x \to a} u(x)\right]^n = A^n$

5) $\lim\limits_{x \to a} u(x) = \lim\limits_{x \to a} v(x)$ if $u(x) = v(x)$ $(x \neq a)$

6) $\lim\limits_{x \to \infty} \dfrac{1}{x^n} = 0$ and $\lim\limits_{x \to -\infty} \dfrac{1}{x^n} = 0$, n is a positive integer

where

> $a, k, h, n, A,$ and B are real numbers.

3. Slope of Tangent Line

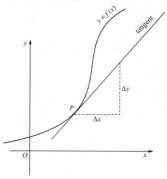

The gradient m of a curve $y = f(x)$ varies from point to point. The gradient of a curve is the slope of the tangent at some point P on a curve $y = f(x)$:

$$m = \frac{\Delta y}{\Delta x}$$

CALCULUS
Derivatives

---

4. Definition of the Derivative

For any function $y = f(x)$, between points P and P_1,

$$\frac{\Delta y}{\Delta x} = \frac{f(x + \Delta x) - f(x)}{\Delta x}$$

is the average rate of change of the function $y = f(x)$, and its limit as Δx approaches zero is the derivative of the function $y = f(x)$. The process of finding this limit, the derivative, is called *differentiation*.

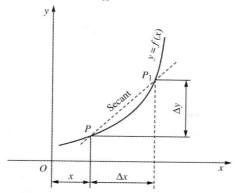

The derivative of the function $y = f(x)$ may be denoted in any of the following ways,

$$f'(x), \quad y', \quad \frac{dy}{dx}, \text{ or } \frac{d}{dx}[f(x)]$$

Hence,

$$y' = \lim_{\Delta x \to 0} \frac{\Delta y}{\Delta x} = \lim_{\Delta x \to 0} \frac{f(x + \Delta x) - f(x)}{\Delta x}$$

CALCULUS
Derivatives

---

5. Basic Derivatives

Function	Derivative
Basic rules	
$y = k,\quad k$ is a real number	$y' = 0$
$y = c \cdot x^n + C$	$y' = c \cdot n \cdot x^{n-1}$
$y = u(x) \pm v(x)$	$y' = u'(x) \pm v'(x)$
$y = u(x) \cdot v(x)$	$y' = u' \cdot v + u \cdot v'$
$y = \dfrac{u(x)}{v(x)} \quad v(x) \neq 0$	$y' = \dfrac{u' \cdot v - u \cdot v'}{v^2}$
$y = \sqrt{x}$	$y' = \dfrac{1}{2\sqrt{x}}$
Chain Rule	
$y = f[u(x)]$	$y' = f'(u) \cdot u'(x)$ or $\dfrac{dy}{dx} = \dfrac{dy}{du} \cdot \dfrac{du}{dx}$
Parametric Form of Derivative	
$y = f(x) \begin{cases} x = f_1(t) \\ y = f_{22}(t) \end{cases}$	$y' = \dfrac{dy}{dt} \cdot \dfrac{dt}{dx}$
Derivatives of Exponential Functions	
$y = e^x$	$y' = e^x = y''; \quad y'' = \dfrac{d^2 y}{dx^2}$

CALCULUS
Derivatives

Continued from # 5

$y = e^{-x}$	$y' = -e^{-x}$
$y = e^{ax}$	$y' = a \cdot e^{ax}$
$y = x \cdot e^{x}$	$y' = e^{x}(1+x)$
$y = \sqrt{e^{x}}$	$y' = \dfrac{\sqrt{e^{x}}}{2}$
$y = a^{x}$	$y' = a^{x}\ln a$
$y = a^{nx}$	$y' = n \cdot a^{nx}\ln a$
$y = a^{x^{2}}$	$y' = a^{x^{2}} \cdot 2x\ln a$
Derivatives of Trigonometric Functions	
$y = \sin x$ $y = \cos x$	$y' = \cos x$ $y' = -\sin x$
$y = \tan x$	$y' = \dfrac{1}{\cos^{2} x} = 1 + \tan^{2} x$
$y = \cot x$	$y' = \dfrac{-1}{\sin^{2} x} = -(1 + \cot^{2} x)$
$y = a \cdot \sin(kx)$	$y' = a \cdot k \cdot \cos(kx)$
$y = a \cdot \cos(kx)$	$y' = -a \cdot k \cdot \sin(kx)$
$y = \sin^{n} x$	$y' = n \cdot \sin^{n-1} x \cdot \cos x$

CALCULUS
Derivatives

Continued from # 5

$y = \cos^n x$	$y' = -n\cos^{n-1} x \sin x$
$y = \tan^n x$	$y' = n\tan^{n-1} x\left(1 + \tan^2 x\right)$
$y = \cot^n x$	$y' = -n \cdot \cot^{n-1} x \cdot \left(1 + \cot^2 x\right)$
$y = \dfrac{1}{\sin x}$	$y' = \dfrac{-\cos x}{\sin^2 x}$
$y = \dfrac{1}{\cos x}$	$y' = \dfrac{\sin x}{\cos^2 x}$

Derivatives of Inverse Trigonometric Functions	
$y = \arcsin x$	$y' = \dfrac{1}{\sqrt{1 - x^2}}$
$y = \arccos x$	$y' = -\dfrac{1}{\sqrt{1 - x^2}}$
$y = \arctan x$	$y' = \dfrac{1}{1 + x^2}$
$y = \operatorname{arc} \cot x$	$y' = -\dfrac{1}{1 + x^2}$
$y = \operatorname{arcsinh} x$	$y' = \dfrac{1}{\sqrt{x^2 + 1}}$

CALCULUS
Derivatives

Continued from # 5

$y = \text{arccosh } x$	$y' = \dfrac{1}{\sqrt{x^2 - 1}}$
$y = \text{arctanh } x$	$y' = \dfrac{1}{1 - x^2}$
$y = \text{arccoth } x$	$y' = \dfrac{1}{1 - x^2}$
Derivatives of Hyperbolic Functions	
$y = \sinh x$	$y' = \cosh x$
$y = \cosh x$	$y' = \sinh x$
$y = \tanh x$	$y' = \dfrac{1}{\cosh^2 x}$
$y = \coth x$	$y' = -\dfrac{1}{\sinh^2 x}$
Derivatives of Logarithmic Functions	
$y = \ln x$	$y' = \dfrac{1}{x},$
$y = \log_a x$	$y' = \dfrac{1}{x \cdot \ln a}$
$y = \ln(1 \pm x)$	$y' = \pm \dfrac{1}{1 \pm x}$

CALCULUS
Derivatives

Continued from # 5

$y = \ln x^n$	$y' = \dfrac{n}{x}$
$y = \ln \sqrt{x}$	$y' = \dfrac{1}{2x}$

6. Increasing and Decreasing Functions

If $y'(x) > 0$, function $y(x)$ increases for each value of x in an interval (a, b).

If $y'(x) < 0$, function $y(x)$ decreases for each value of x in an interval (a, b).

If $y'(x) = 0$, function $y(x)$ is tangentially parallel to the x-axis at x.

7. Maximum and Minimum of a Function

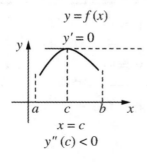

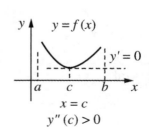

If $y''(c) > 0$ and $y'(c) = 0$, there is a minimum at $x = c$

If $y''(c) < 0$ and $y'(c) = 0$, there is a maximum at $x = c$.

If $y''(c) = 0$, then the test gives no information.

8. Solving Applied Problems

Step 1: Read problem carefully.

Step 2: If possible, sketch a diagram.

Step 3: Decide on the variable whose values must be maximized or minimized. Express that variable as a function of one other variable.

Step 4: Find the critical points for the function of Step 3. Check if each is a maximum or minimum.

Step 5: Check the extrema at any end point of the domain of the function of Step 3.

Step 6: Check to be sure the answer is reasonable.

9. Integration

Integration is the opposite of derivation In calculus integration of a given real valid function $y = f(x)$ is a function $F(x)$ whose derivative is equal to $f(x)$, i.e.,

$$F'(x) = \frac{dF(x)}{dx} = f(x)$$

CALCULUS
Integration

There are two meanings of integration: definite integrals and indefinite integrals.

a) Indefinite integrals
The integral of a function is a special limit with many diverse applications.

If $$F'(x) = f(x), \text{ then}$$

$$\int f(x)dx = F(x) + C$$

where
C = unknown constant

b) Definite integral
If $f(x)$ is continuous on the interval $[a, b]$, the definite integral of $f(x)$ from a to b is given by

$$\int_a^b f(x)dx = F(x)\Big|_a^b = F(b) - F(a)$$

10. Basic Integration Rules
1) The indefinite integral of a constant
$$\int k \, dx = kx + C \quad (k = \text{constant})$$
2) The power rule for indefinite integrals

CALCULUS
Integration

$$\int x^n dx = \frac{1}{n+1} x^{n+1} + C$$

3) The indefinite integral of a constant multiple of a function

$$\int c \cdot f(x) dx = c \int f(x) dx \quad (c = \text{constant})$$

4) The sum rule

$$\int [f(x) \pm g(x)] dx = \int f(x) dx \pm \int g(x) dx$$

5) The indefinite integral of the exponential function

$$\int e^x dx + C$$

6) The indefinite integral of the function $f(x) = x^{-1}$

$$\int x^{-1} dx = \int \frac{1}{x} dx = \ln|x| + C \quad (x \neq 0)$$

11. Integration by Substitution

The method of substitution is related to the chain rule for differentiating functions.

There are five steps involved in integration by substitution.

Consider the indefinite integral

$$\int f[g(x)] g'(x) dx$$

CALCULUS
Integration

Step 1: Let $u = g(x)$, where $g(x)$ is part of the integrand, usually the inside function of the composite function $f[g(x)]$.

Step 2: Determine $g'(x) = \dfrac{du}{dx}$

Step 3: Use the substitutions $u = g(x)$ and $du = g'(x)dx$ to convert the entire integral into one involving only u.

Step 4: Evaluate the resulting integral.

Step 5: Replace u with $g(x)$ to obtain the final solution as a function of x.

Example:
    Find
$$F(x) = \int \sqrt{3x - 5}\,dx$$

Solution:

Step 1: Observe that the integrand involves the composite function $\sqrt{3x - 5}$ with the "inside function" $g(x) = 3x - 5$. So, we choose
$$u = 3x - 5$$

Step 2: Compute $du = g'(x) = 3dx$

Step 3: Making the substitution $u = 3x - 5$ and $du = g'(x) = 3dx$, we obtain

$$F(x) = \frac{1}{3}\int \sqrt{u}\,du$$

Step 4: Evaluate

$$F(x) = \frac{1}{3}\int \sqrt{u}\,du = \frac{2}{9}u\sqrt{u} + C$$

Step 5: Replacing u with $3x-5$ we obtain

$$F(x) = \frac{2}{9}(3x - 5)\sqrt{3x - 5} + C$$

12. Basic Integrals

1) $\int \dfrac{dx}{x^n} = -\dfrac{1}{n-1} \cdot \dfrac{1}{x^{n-1}} + C \quad (n \neq 1)$

2) $\int a^{bx}\,dx = \dfrac{1}{b} \cdot \dfrac{a^{bx}}{\ln|a|} + C$

3) $\int (\ln x)^2\,dx = x(\ln|x|)^2 - 2x\ln|x| + 2x + C$

4) $\int \dfrac{dx}{\ln x} = \ln\big|(\ln|x|)\big| + \ln|x| + \dfrac{(\ln|x|)^2}{2 \cdot 2!} + \dfrac{(\ln|x|)^3}{3 \cdot 3!} + \cdots$

5) $\int x\ln x\,dx = x^2\left[\dfrac{\ln|x|}{2} - \dfrac{1}{4}\right] + C$

CALCULUS
Basic Integrals

6) $\int \dfrac{dx}{x \ln x} = \ln\left|\left(\ln|x|\right)\right| + C$

7) $\int e^{ax}\, dx = \dfrac{1}{a} e^{ax} + C$

8) $\int x e^{ax}\, dx = \dfrac{e^{ax}}{a^2}(ax - 1) + C$

9) $\int x^2 e^{ax}\, dx = e^{ax}\left(\dfrac{x^2}{a} - \dfrac{2x}{a^2} + \dfrac{2}{a^3}\right) + C$

10) $\int x^n e^{ax}\, dx = \dfrac{1}{a} x^n e^{ax} - \dfrac{n}{a}\int x^{n-1} e^{ax}\, dx + C$

11) $\int \dfrac{dx}{1 + e^{ax}} = \dfrac{1}{a}\ln\left|\dfrac{e^{ax}}{1 + e^{ax}}\right| + C$

12) $\int \dfrac{e^{ax}\, dx}{b + c e^{ax}} = \dfrac{1}{ac}\ln\left|b + c e^{ax}\right| + C$

13) $\int e^{ax} \ln x\, dx = \dfrac{e^{ax} \ln|x|}{a} - \dfrac{1}{a}\int \dfrac{e^{ax}}{x}\, dx + C$

14) $\int e^{ax} \cos bx\, dx = \dfrac{e^{ax}}{a^2 + b^2}(a \cos bx + b \sin bx) + C$

15) $\int e^{ax} \sin bx\, dx = \dfrac{e^{ax}}{a^2 + b^2} + (a \sin bx - b \cos bx) + C$

16) $\int \dfrac{dx}{ax + b} = \dfrac{1}{a}\ln|ax + b| + C$

CALCULUS
Basic Integrals

17) $\int \dfrac{dx}{(ax+b)^n} = -\dfrac{1}{a(n-1)(ax+b)^{n-1}} + C \quad (n \neq 1)$

18) $\int \dfrac{dx}{ax-b} = \dfrac{1}{a}\ln|ax-b| + C$

19) $\int \dfrac{dx}{(ax-b)^n} = -\dfrac{1}{a(n-1)(ax-b)^{n-1}} + C \quad (n \neq 1)$

20) $\int \dfrac{dx}{(ax+b)(cx+d)} = \dfrac{1}{bc-ad}\ln\left|\dfrac{cx+d}{ax+b}\right| + C \,(bc-ad \neq 0)$

21) $\int \dfrac{dx}{(ax-b)(cx-d)} = \dfrac{1}{ad-bc}\ln\left|\dfrac{cx-d}{ax-b}\right| + C,$
$$(bc-ad \neq 0)$$

22) $\int \dfrac{x\,dx}{ax+b} = \dfrac{x}{a} - \dfrac{b}{a^2}\ln|ax+b| + C$

23) $\int \dfrac{x^2\,dx}{ax+b} = \dfrac{1}{a^3}\left[\begin{array}{l}\dfrac{1}{2}(ax+b)^2 - 2b(ax+b) + \\ + b^2\ln|ax+b|\end{array}\right] + C$

24) $\int \dfrac{dx}{x(ax+b)} = -\dfrac{1}{b}\ln\left|a + \dfrac{b}{x}\right| + C$

25) $\int \dfrac{x^3\,dx}{ax+b} = \dfrac{1}{a^4}\left[\begin{array}{l}\dfrac{1}{3}(ax+b)^3 - \dfrac{3}{2}b(ax+b)^2 + \\ + 3b^2(ax+b) - b^3\ln|ax+b|\end{array}\right] + C$

CALCULUS
Basic Integrals

---

26) $\int \dfrac{dx}{x^2(ax+b)} = -\dfrac{1}{bx} + \dfrac{a}{b^2} \ln\left|a + \dfrac{b}{x}\right| + C$

27) $\int \dfrac{dx}{a^2+x^2} = \dfrac{1}{a} \arctan \dfrac{x}{a} + C$

28) $\int \dfrac{x\,dx}{a^2+x^2} = \dfrac{1}{2} \ln\left|a^2+x^2\right| + C$

29) $\int \dfrac{x^2\,dx}{a^2+x^2} = x - a \arctan \dfrac{x}{a} + C$

30) $\int \dfrac{x^3\,dx}{a^2+x^2} = \dfrac{x^2}{2} - \dfrac{a^2}{2} \ln\left|a^2+x^2\right| + C$

31) $\int \dfrac{dx}{a^2-x^2} = -\int \dfrac{dx}{x^2-a^2} = \dfrac{1}{2a} \ln\left|\dfrac{a+x}{a-x}\right| + C$

32) $\int \dfrac{x\,dx}{a^2-x^2} = -\int \dfrac{x\,dx}{x^2-a^2} = -\dfrac{1}{2} \ln\left|a^2-x^2\right| + C$

33) $\int \dfrac{x^2\,dx}{a^2-x^2} = -\int \dfrac{x^2\,dx}{x^2-a^2} = -x + a\dfrac{1}{2} \ln\left|\dfrac{a+x}{a-x}\right| + C$

34) $\int \dfrac{x^3\,dx}{a^2-x^2} = -\int \dfrac{x^3\,dx}{x^2-a^2} = -\dfrac{x^2}{2} - \dfrac{a^2}{2} \ln\left|a^2-x^2\right| + C$

35) $\int \dfrac{x\,dx}{\left(a^2+x^2\right)^2} = -\dfrac{1}{2\left(a^2+x^2\right)} + C$

36) $\int \dfrac{x^2\,dx}{\left(a^2+x^2\right)^2} = -\dfrac{x}{2\left(a^2+x^2\right)} + \dfrac{1}{2a} \arctan \dfrac{x}{a} + C$

CALCULUS
Basic Integrals

---

37) $\int \dfrac{x^3 dx}{\left(a^2 + x^2\right)^2} = \dfrac{a^2}{2\left(a^2 + x^2\right)} + \dfrac{1}{2}\ln\left|a^2 + x^2\right| + C$

38) $\int \dfrac{dx}{\left(a^2 - x^2\right)^2} = \dfrac{x}{2a^2\left(a^2 - x^2\right)} + \dfrac{1}{2a^3}\cdot\dfrac{1}{2}\ln\left|\dfrac{a+x}{a-x}\right| + C$

39) $\int \dfrac{xdx}{\left(a^2 - x^2\right)^2} = \dfrac{1}{2\left(a^2 - x^2\right)} + C$

40) $\int \dfrac{x^2 dx}{\left(a^2 - x^2\right)^2} = \dfrac{x}{2\left(a^2 - x^2\right)} - \dfrac{1}{2a}\cdot\dfrac{1}{2}\ln\left|\dfrac{a+x}{a-x}\right| + C$

41) $\int \dfrac{x^3 dx}{\left(a^2 - x^2\right)^2} = \dfrac{a^2}{2\left(a^2 - x^2\right)} + \dfrac{1}{2}\ln\left|a^2 - x^2\right| + C$

42) $\int \sqrt{x}\,dx = \dfrac{2}{3}\sqrt{x^3} + C$

43) $\int \sqrt{ax + b}\,dx = \dfrac{2}{3a}\sqrt{(ax + b)^3} + C$

44) $\int x\sqrt{ax + b}\,dx = \dfrac{2(3ax - 2b)\sqrt{(ax + b)^3}}{15a^2} + C$

45) $\int \dfrac{dx}{\sqrt{x}} = 2\sqrt{x} + C$

46) $\int \dfrac{dx}{\sqrt{ax + b}} = \dfrac{2\sqrt{(ax + b)}}{a} + C$

47) $\int \dfrac{xdx}{\sqrt{ax + b}} = \dfrac{2(ax - 2b)\sqrt{(ax + b)}}{3a^2} + C$

CALCULUS
Basic Integrals

---

48) $\int \dfrac{x^2 dx}{\sqrt{ax+b}} = \dfrac{2\left(3a^2 x^2 - 4abx + 8b^2\right)\sqrt{(ax+b)}}{15a^3} + C$

49) $\int \sqrt{a^2 + x^2}\,dx = \dfrac{x}{2}\sqrt{a^2 + x^2} + \dfrac{a^2}{2} \operatorname{arcsinh}\dfrac{x}{a} + C$

50) $\int x\sqrt{a^2 + x^2}\,dx = \dfrac{1}{3}\sqrt{(a^2 + x^2)^3} + C$

51) $\int x^3\sqrt{a^2 + x^2}\,dx = \dfrac{\sqrt{(a^2 + x^2)^5}}{5} - \dfrac{a^2\sqrt{(a^2 + x^2)^3}}{3} + C$

52) $\int x^2\sqrt{a^2 + x^2}\,dx = \dfrac{x}{4}\sqrt{(a^2 + x^2)^3}$

$\qquad - \dfrac{a^2}{8}\left(x\sqrt{a^2 + x^2}\right) - \dfrac{a^4}{8}\operatorname{arcsinh}\dfrac{x}{a} + C$

53) $\int \dfrac{\sqrt{a^2 + x^2}}{x}\,dx = \sqrt{a^2 + x^2} - a\ln\left|\dfrac{a + \sqrt{a^2 + x^2}}{x}\right| + C$

54) $\int \dfrac{\sqrt{a^2 + x^2}}{x^2}\,dx = -\dfrac{\sqrt{a^2 + x^2}}{x} + \operatorname{arcsinh}\dfrac{x}{a} + C$

55) $\int \dfrac{dx}{\sqrt{a^2 + x^2}} = \operatorname{arcsinh}\dfrac{x}{a} + C$

56) $\int \dfrac{x\,dx}{\sqrt{a^2 + x^2}} = \sqrt{a^2 + x^2} + C$

57) $\int \dfrac{x^2 dx}{\sqrt{a^2 + x^2}} = \dfrac{x}{2}\sqrt{a^2 + x^2} - \dfrac{a^2}{2}\operatorname{arcsinh}\dfrac{x}{a} + C$

CALCULUS
Basic Integrals

---

58) $\int \dfrac{x^3 dx}{\sqrt{a^2 + x^2}} = \dfrac{\sqrt{(a^2 + x^2)^3}}{3} - a^2 \sqrt{x^2 + a^2} + C$

59) $\int \dfrac{dx}{x\sqrt{a^2 + x^2}} = -\dfrac{1}{a} \ln\left|\dfrac{a + \sqrt{a^2 + x^2}}{x}\right| + C$

60) $\int \dfrac{dx}{x^2\sqrt{a^2 + x^2}} = -\dfrac{\sqrt{x^2 + a^2}}{a^2 x} + C$

61) $\int x\sqrt{a^2 - x^2}\, dx = -\dfrac{1}{3}\sqrt{(a^2 - x^2)^3} + C$

62) $\int \dfrac{dx}{\sqrt{a^2 - x^2}} = \arcsin\dfrac{x}{a} + C$

63) $\int \dfrac{x dx}{\sqrt{a^2 - x^2}} = -\sqrt{a^2 - x^2} + C$

64) $\int \dfrac{x^2 dx}{\sqrt{a^2 - x^2}} = -\dfrac{x}{2}\sqrt{a^2 - x^2} + \dfrac{a^2}{2}\arcsin\dfrac{x}{a} + C$

65) $\int \dfrac{x^3 dx}{\sqrt{a^2 - x^2}} = \dfrac{\sqrt{(a^2 - x^2)^3}}{3} - a^2 \sqrt{a^2 - x^2} + C$

66) $\int \dfrac{dx}{x\sqrt{a^2 - x^2}} = -\dfrac{1}{a} \ln\left|\dfrac{a + \sqrt{a^2 - x^2}}{x}\right| + C$

67) $\int \dfrac{dx}{x^2\sqrt{a^2 - x^2}} = -\dfrac{\sqrt{a^2 - x^2}}{a^2 x} + C$

CALCULUS
Basic Integrals

---

68) $\int \dfrac{\sqrt{a^2 - x^2}}{x} dx = \sqrt{a^2 - x^2} - a \operatorname{arccosh} \dfrac{a}{x} + C$

69) $\int \dfrac{\sqrt{a^2 - x^2}}{x^2} dx = -\dfrac{\sqrt{a^2 - x^2}}{x} + \operatorname{arccos} \dfrac{x}{a} + C$

70) $\int \dfrac{\sqrt{a^2 - x^2}}{x^3} dx = -\dfrac{\sqrt{a^2 - x^2}}{2x^2} + \dfrac{1}{2a} \operatorname{arccosh} \dfrac{a}{x} + C$

71) $\int \cos x dx = \sin x + C$

72) $\int \sin x dx = -\cos x + C$

73) $\int \tan x dx = \ln|\sec x| + C$

74) $\int \cot x dx = \ln|\sin x| + C$

75) $\int \sec x dx = \ln|\sec x + \tan x| + C$

76) $\int \csc x dx = \ln|\csc x - \cot x| + C$

77) $\int \csc^2 x dx = -\cot x + C$

78) $\int \sec x \tan x dx = \sec x + C$

79) $\int \csc x \cot x dx = -\csc x + C$

80) $\int \sin(bx) dx = -\dfrac{1}{b} \cos(bx) + C$

81) $\int \sin^2(bx) dx = \dfrac{x}{2} - \dfrac{1}{4b} \sin(2bx) + C$

82) $\int \cos(bx) dx = \dfrac{1}{b} \sin(bx) + C$

CALCULUS
Basic Integrals

---

83) $\int \tan(bx)dx = -\dfrac{1}{b}\ln|\cos(bx)| + C$

84) $\int \tan^2(bx)dx = \dfrac{1}{b}\tan bx - x + C$

85) $\int \sin(ax)\sin(bx)dx = -\dfrac{\sin(ax+bx)}{2(a+b)} + \dfrac{\sin(ax-bx)}{2(a-b)} + C$

$$(|a| \neq |b|).$$

86) $\int \cos(ax)\cos(bx)dx = \dfrac{\sin(ax+bx)}{2(a+b)} + \dfrac{\sin(ax-bx)}{2(a-b)} + C$

$$(|a| \neq |b|)$$

87) $\int \sin(ax)\cos(bx)dx = -\dfrac{\cos(ax+bx)}{2(a+b)} - \dfrac{\cos(ax-bx)}{2(a-b)} + C$

88) $\int x^n \sin bx\,dx = -\dfrac{x^n}{b}\cos bx + \dfrac{n}{b}\int x^{n-1}\cos bx\,dx + C$

89) $\int x^n \cos bx\,dx = \dfrac{x^n}{b}\sin bx - \dfrac{n}{b}\int x^{n-1}\sin bx\,dx + C$

90) $\int \sin^n x\,dx = -\dfrac{1}{n}\sin^{n-1} x \cos x + \dfrac{n-1}{n}\int \sin^{n-2} x\,dx + C$

91) $\int \cos^n x\,dx = \dfrac{1}{n}\cos^{n-1} x \sin x + \dfrac{n-1}{n}\int \cos^{n-2} x\,dx + C$

92) $\int \arcsin x\,dx = x\arcsin x + \sqrt{1-x^2} + C$

93) $\int \arccos x\,dx = x\arccos x - \sqrt{1-x^2} + C$

CALCULUS
Basic Integrals

94) $\int \arctan x \, dx = x \arctan x - \dfrac{1}{2} \ln\left|1 + x^2\right| + C$

95) $\int \text{arc cot } x \, dx = x \text{ arccot } x + \dfrac{1}{2} \ln\left|1 + x^2\right| + C$

96) $\int \sinh(ax) \, dx = \dfrac{1}{a} \cosh(ax) + C$

97) $\int \sinh^2 x \, dx = \dfrac{1}{4} \sinh(2x) - \dfrac{x}{2} + C$

98) $\int \cosh(ax) \, dx = \dfrac{1}{a} \sinh(ax) + C$

99) $\int \cosh^2 x \, dx = \dfrac{1}{4} \sinh(2x) + \dfrac{x}{2} + C$

100) $\int \cosh^n x \, dx = \dfrac{1}{n} \sinh x \cosh^{n-1} x$

$\qquad + \dfrac{n-1}{n} \int \cosh^{n-2} x \, dx + C \quad (n > 0)$

101) $\int \tanh(ax) \, dx = \dfrac{1}{a} \ln\left|\cosh(ax)\right| + C$

102) $\int \tanh^2 x \, dx + x - \tanh x + C$

103) $\int \tanh^n x \, dx = -\dfrac{1}{n-1} \tanh^{n-1} x$

$\qquad + \int \tanh^{n-2} x \, dx + C \qquad (n \neq 1)$

CALCULUS
Applications of Integration

104) $\int \coth(ax)dx = \dfrac{1}{a}\ln|\sinh(ax)| + C$

105) $\int \coth^2 x\,dx = x - \coth x + C$

106) $\int \dfrac{dx}{\sinh ax} = \dfrac{1}{a}\ln\left|\tanh\dfrac{ax}{2}\right| + C$

107) $\int \dfrac{dx}{\sinh^2 x} = -\coth x + C$

108) $\int \dfrac{dx}{\cosh ax} = \dfrac{2}{a}\arctan e^{ax} + C$

109) $\int \dfrac{dx}{\cosh^2 x} = \tanh x + C$

13. Arc Length

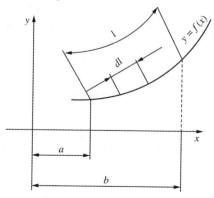

CALCULUS
Applications of Integration

---

a) Arc differential:

$$dl = \sqrt{dx^2 + dy^2} = \sqrt{1 + \left(\frac{dy}{dx}\right)^2}\, dx$$

b) Arc length

Length of curve $y = f(x)$ from $x = a$ to $x = b$ is

$$l = \int_a^b \sqrt{1 + y'^2}\, dx; \quad y' = \frac{dy}{dx}$$

c) Surface area of solid of revolution

Surface area obetained when the curve $y = f(x)$ is rotated around the x-axis is

$$A = 2\pi \int_a^b y\sqrt{1 + y'^2}\, dx$$

14. Finding an Area and a Volume

a) Area

Area A below the curve $y = f(x)$ from $x = a$ to $x = b$ is

$$A = \int_a^b y\, dx$$

b) Volume

1) Volume of the body obtained when region A is rotated around the x-axis:

CALCULUS
Applications of Integration

$$V = \pi \int_a^b y^2\, dx$$

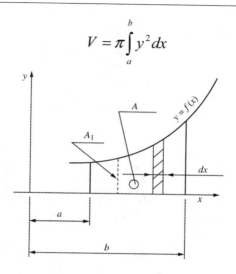

2) Volume of a body the cross-section A_1 of which is a function of x:

$$V = \int_a^b A_1\, dx$$

15. Finding the Area between Two Curves

Area A between curve $y = f(x)$ and $y = g(x)$ from $x = a$ to $x = b$ is

$$A = \int_a^b f(x)\,dx - \int_a^b g(x)\,dx = \int_a^b [f(x) - g(x)]\,dx$$

CALCULUS
Applications of Integration

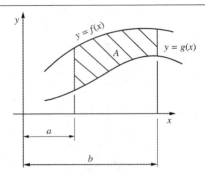

Example:

Find the area of the region A bounded by the graphs of $f(x) = 2x - 1$ and $g(x) = x^2 - 3$ and the vertical lines $x = 0$ and $x = 2$.

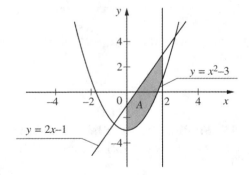

Solution:

$$\int_0^2 [f(x) - g(x)]dx = \int_0^2 [(2x - 1) - (x^2 - 3)]dx$$

CALCULUS
Applications of Integration

$$= \int_0^2 \left(-x^2 + 2x + 2\right)dx$$

$$= -\frac{1}{3}x^3 + x^2 + 2x \Big|_0^2$$

$$= \left(-\frac{8}{3} + 4 + 4\right) - 0 = \frac{16}{3}$$

Hence, area $A = \dfrac{16}{3} = 5\dfrac{1}{3}$

STATISTICS

Statistics is the mathematics of the collection, organization, and interpretation of numerical data, especially the analysis of population characteristics by inference from sampling. The most familiar statistical measure is the arithmetic mean, which is an average value for a group of numerical observations.

A second important statistic or statistical measure is the standard deviation, which is a measure of how much the individual observations are scattered about the mean.

This section contains the most frequently used formulas, rules, and definitions regarding the following:

- Sets
- Permutations and Combinations
- Probability
- Distributions
- Reliability

STATISTICS
Sets

---

1. Definition of Set and Notation

A set is a collection of objects called elements. In mathematics we write a set by putting its elements between curly brackets $\{\ \}$.

Set A, which contains numbers 3, 4, and 5, is written

$$A = \{3, 4, 5\}$$

a) Empty set

A set with no elements is called an empty set and is denoted by

$$\{\ \} = \varnothing$$

b) Subset

If every element of one set also belongs to another set, for example

$$A = \{3, 4, 5\} \text{ and } B = \{1, 2, 3, 4, 5, 6, 7\},$$

then set A is a subset of set B because every element of set A is also an element of set B, and is written as

$$A \subseteq B$$

c) Set equality

The sets A and B are equal if and only if they have exactly the same elements, and the equality is written as

$$A = B$$

---

d) Set union

The union of set A and set B is the set of all elements that belong to either A or B or both, and is written as

$$A \cup B = \left\{x \middle| x \in A \text{ or } x \in B \text{ or } \text{both}\right\}$$

2. Terms and Symbols

$\{\ \}$ set braces

$\in$ is an element of

$\notin$ is not an element of

$\subseteq$ is a subset of

$\not\subset$ is not a subset of

A' complement of set A

$\cap$ set intersection

$\cup$ set union

3. Venn Diagrams

Venn diagrams are used to visually illustrate relationships between sets.

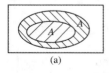
(a)

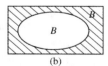
(b)

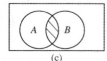
(c)

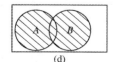
(d)

These Venn diagrams illustrate the following statements:

   a) Set A is a subset of set B ($A \subset B$).
   b) Set B' is the complement of set B.
   c) Two sets A and B with their intersection $A \cap B$.
   d) Two sets A and B with their union $A \cup B$.

4. Operations on Sets

If A, B and C are arbitrary subsets of universal set U, then the following rules govern the operations on sets:

1) Commutative law for union

$$A \cup B = B \cup A$$

2) Commutative law for intersection

$$A \cap B = B \cap A$$

3) Associative law for union

$$A \cup (B \cup C) = (A \cup B) \cup C$$

4) Associative law for intersection

$$A \cap (B \cap C) = (A \cap B) \cap C$$

5) Distributive law for union

$$A \cup (B \cap C) = (A \cup B) \cap (A \cup C)$$

6) Distributive law for intersection

$$A \cap (B \cup C) = (A \cap B) \cup (A \cap C)$$

5. De Morgan's Laws

$$(A \cup B)' = A' \cap B' \quad (1)$$
$$(A \cap B)' = A' \cup B' \quad (2)$$

The complement of the union of two sets is equal to the intersection of their complements (equation 1).
The complement of the intersection of two sets is equal to the union of their complements (equation 2).

6. Counting the Elements in a Set
The number of elements in a finite set is determined by simply counting the elements in the set.

If A and B are disjoint sets, then

$$n(A \cup B) = n(A) + n(B)$$

In general, A and B need not be disjoint, so

STATISTICS
Permutations and Combinations

$$n(A \cup B) = n(A) + n(B) - n(A \cap B)$$

where

n = number of the elements in a set

7. Permutations

A permutation of m elements from a set of n elements is any arrangement, without repetition, of the m elements. The total number of all the possible permutations of m distinct objects taken from a set of n elements is

$$P(n,m) = \frac{n!}{(n-m)!} \qquad (n \geq m)$$

Example:

Find the number of ways a president, vice-president, secretary, and treasurer can be chosen from a committee of eight members.

Solution:

$$P(n,m) = \frac{n!}{(n-m)!} = P(8,4) = \frac{8!}{(8-4)!}$$
$$= \frac{8 \cdot 7 \cdot 6 \cdot 5 \cdot 4 \cdot 3 \cdot 2 \cdot 1}{4 \cdot 3 \cdot 2 \cdot 1} = 1680$$

There are 1,680 ways of choosing the four officials from the committee of eight members.

STATISTICS
Permutations and Combinations

---

8. Combinations

The number of combinations of m distinct elements taken from a set of n elements is given by

$$C(n,m) = \frac{n!}{m!(n-m)!} \qquad (n \geq m)$$

Example:

How many poker hands of five cards can be dealt from a standard deck of 52 cards?

Solution:

Note: The order in which the 5 cards are dealt is not important.

$$C(n,m) = \frac{n!}{m!(n-m)!} = C(52,5) = \frac{52!}{5!(52-5)!} = \frac{52!}{5!\,47!}$$

$$= \frac{52 \cdot 51 \cdot 50 \cdot 49 \cdot 48}{5 \cdot 4 \cdot 3 \cdot 2 \cdot 1} = 2,598,960$$

9. Probability Terminology

A number of specialized terms are used in the study of probability.

Experiment: An experiment is an activity or occurrence with an observable result.

Outcome: The outcome is the result of the experiment.

STATISTICS
Probability

Sample point: This is an outcome of an experiment.

Event: An event is a set of outcomes (a subset of the sample space) to which a probability is assigned.

10. Basic Probability Principles

Consider a *random sampling process* in which all the outcomes solely depend on *chance*, i.e., each outcome is equally likely to happen. If S is a uniform sample space and the collection of desired outcomes is E, the probability of the desired outcomes is

$$P(E) = \frac{n(E)}{n(S)}$$

where

$n(E)$ = number of favorable outcomes in E
$n(S)$ = number of possible outcomes in S

Since E is a subset of S,

$$0 \le n(E) \le n(S),$$

so the probability of the desired outcome is

$$0 \le P(E) \le 1$$

---

11. Random Variable

A random variable is a quantity that is assigned a number according to the outcome of a chance experiment.

Example:

1. A coin is tossed six times. The random variable X is the number of tails that are noted. X can only take the values 1, 2,…, 6, so X is a discrete random variable.

2. A light bulb is used until it burns out. The random variable Y is its lifetime in hours. Y can take any positivereal value, so Y is a continuous random variable.

12. Mean Value $\overline{x}$ or Expected Value μ

The mean value or expected value of a random variable indicates its average or central value. It is a useful summary value of the variable's distribution.

1) If the random variable X is discrete, then its mean value is

$$\overline{x} = x_1 p_1 + x_2 p_2 + \ldots + x_n p_n = \sum_{i=1}^{n} x_i p_i$$

where

p_i = probability that X takes the value x_i

STATISTICS
Distributions

---

2) If X is a continuous random variable with probability density function $f(x)$, then the expected value of X is

$$\mu = E(X) = \int_{-\infty}^{+\infty} xf(x)dx$$

where

$f(x) =$ probability density function

13. Variance

The variance is a measure of the "spread" of a random variable's distribution about its average value.

a) Discrete system:

$$\sigma^2 = \sum_{i=1}^{n} \left(x_i - \bar{x} \right)^2 p_i$$

b) Continuous system:

$$\sigma^2 = \int_{-\infty}^{+\infty} (x - \mu)^2 \cdot f(x)dx$$

14. Standard Deviation

Standard deviation, denoted by σ is the positive square root of the variance. Both variance and standard

STATISTICS
Distributions

---

deviation are used to describe the spread of a distribution.

a) Discrete system:

$$\sigma = \sqrt{\sum_{i=1}^{n}\left(x_i - \bar{x}\right)^2 p_i}$$

b) Continuous system:

$$\sigma = \sqrt{\int_{-\infty}^{+\infty}(x - \mu)^2 \cdot f(x)dx}$$

15. Normal Distribution

The normal distribution, or Gaussian distribution, is a symmetrical distribution commonly referred to as a bell curve.

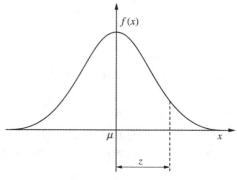

a) Probability density function:

$$f(x) = \frac{1}{\sigma\sqrt{2\pi}} e^{\frac{-(x-\mu)^2}{2\sigma^2}}$$

b) Distribution function:

$$F(x) = \int_{-\infty}^{x} \frac{1}{\sigma\sqrt{2\pi}} e^{\frac{-(t-\mu)^2}{2\sigma^2}} \, dt$$

c) Standard value (z-score)
If normal distribution has mean μ and standard deviation σ, then the z-score for the number x is

$$z = \frac{x-\mu}{\sigma}.$$

16. Binomial Distribution

The binomial distribution, also known as the Bernoulli distribution, describes random sampling processes such that all outcomes are either yes or no (success/failure) without ambiguity.

Suppose that the probability of success in a single trial is p in a random sampling process and the failure rate is q, where

$$q = 1 - p$$

STATISTICS
Distributions

The binomial distribution with exactly x successes in n trials, where $x \leq n$, has the following important properties:

a) Density function:

$$f(x) = \frac{n!}{x(n-x)!} p^n q^{n-x}$$

b) Mean:

$$\mu = np$$

c) Variance:

$$\sigma^2 = npq$$

d) Standard deviation:

$$\sigma = \sqrt{npq}$$

17. Poisson Distribution

The Poisson distribution describes a random sampling process in which the desired outcomes occur relatively infrequently but at a regular rate.

Suppose there are on average λ successes in a large number of trials (large sampling period). The Poisson distribution with exactly x successes in the same sampling period has the following important properties.

a) Density function:

$$f(x) = \frac{\lambda^x e^{-\lambda}}{x!}$$

b) Mean:

$$\mu = \lambda = np$$

where n is the (large) number of trials and p is the (small) probability of success on each trial.

c) Distribution function:

$$F(x_j) = \sum_{k \leq j} \frac{\lambda^{x_k} e^{-\lambda}}{x_k}$$

d) Variance:

$$\sigma^2 = \lambda = np$$

e) Standard deviation:

$$\sigma = \sqrt{\lambda} = \sqrt{np} \quad (\lambda = \text{constant} > 0)$$

18. Exponential Distribution

The exponential distribution is used for reliability calculations.

a) Density function:

$$f(x) = \lambda e^{-\lambda x} \qquad (\lambda > 0, \; x \geq 0)$$

---

b) Distribution function:

$$F(x) = 1 - e^{-\lambda x}$$

c) Mean:

$$\mu = \frac{1}{\lambda}$$

d) Variance:

$$\sigma^2 = \frac{1}{\lambda^2}$$

e) Standard deviation:

$$\sigma = \sqrt{\frac{1}{\lambda^2}} = \frac{1}{\lambda}$$

19. General Reliability Definitions

a) Reliability function
The reliability function $R(t)$, also known as the survival function $S(t)$, is defined by

$$R(t) = S(t) = 1 - F(t)$$

b) Failure distribution function
The failure distribution function is the probability of an item failing in the time interval $0 \leq \tau \leq t$:

$$F(t) = \int\limits_0^t f(\tau)\,d\tau \quad (t \geq 0)$$

c) Failure rate

The failure rate of a unit is

$$z(t) = \lim_{\Delta t \to 0} \frac{F(t) - F(t - \Delta t)}{\Delta t\, R(t)} = \frac{f(t)}{R(t)}$$

d) Mean time to failure

The mean time to failure (*MTTF*) of a unit is

$$MTTF = \int\limits_0^\infty f(t) \cdot t\, dt = \int\limits_0^\infty R(t)\, dt$$

e) Reliability of the system

The reliability of a system is the product of the reliability functions of the components $R_1, ..., R_n$

$$R_S(t) = R_1 \cdot R_2 \cdot ... \cdot R_n = \prod_{i=1}^n R_i(t)$$

20. Exponential Distribution Used as Reliability Function

a) Reliability function:

$$R(t) = e^{-\lambda t} \quad (\lambda = \text{constant})$$

STATISTICS
Reliability

---

b) Failure distribution function:

$$F(t) = 1 - e^{-\lambda t}$$

c) Density function of failure:

$$f(t) = \lambda e^{-\lambda t}$$

d) Failure rate:

$$z(t) = \frac{f(t)}{R(t)} = \lambda$$

e) Mean time to failure:

$$MTTF = \int_0^\infty e^{-\lambda t}\, dt = \frac{1}{\lambda}$$

f) System reliability:

$$R_S(t) = e^{-k} \quad (\text{where } k = t \sum_{i=1}^{n} \lambda_i)$$

g) Cumulative failure rate:

$$z_S = \lambda_1 + \lambda_2 + \ldots + \lambda_n = \sum_{i=1}^{n} \lambda_i = \frac{1}{MTBF}$$

where $MTBF$ = mean time between failures.

MATHEMATICAL FUNDAMENTALS OF COMPUTER SCIENCE

Digital computers operate only with numbers. The way the machine operates with these numbers depends on what the numbers represent and in which form they are presented.

In this section we will outline the different ways of representing numbers and other character data types on computers, as well as examples of their applications for solving problems.

This section contains the following:

- Binary Data Representation
- Number Systems
- Binary Logic
- Binary Arithmetic
- Representation of Signed Numbers for Computers
- Information in Binary Form

MATHEMATICAL FUNDAMENTALS OF COMPUTER SCIENCE
Binary Data Representation

---

1. Binary Data Representation

Binary data is a type of data that is represented or displayed in a binary numeral system. Binary data is the only category of data that can be directly understood and executed by a computer. It is numerically represented by a combination of zeros and ones.

a) Bit

Each 1 or 0; YES or NO; ON or OFF in a binary value is called a *bit*, which is short for *binary digit*. The bit is the smallest unit of information.

b) Byte

A collection of 8 bits is called a *byte*. A byte is a very common unit of storage in electronic memory. It is typically the smallest measurement size listed in operating systems. Therefore, each byte can be used to represent $2^8 = 256$ different values.

c) Word

The maximum amount of data (number of bits) that a central processing unit (CPU) can process at once is called a *word*.

Common versions right now are 32-bit words and 64-bit words. The word size in any given computer is fixed. For example, a byte can hold a number from 0 to 255.

MATHEMATICAL FUNDAMENTALS OF COMPUTER SCIENCE
Number Systems

However, a 32-bit integer can range from 0 to 4,294,967,295. An integer is the native "word size" of the system.

Example:
A byte can hold the following set of numbers:

Signed: from -128 to 127, which is the same as from -2^7 to $2^7 - 1$

Unsigned: from 0 to 255, which equals $2^8 - 1$

A 32-bit integer can hold numbers:

Signed: from $-2,147,483,648$ to 2,147,483,647 which are the same as from $-\left(2^{31}\right)$ to $2^{31} - 1$

Unsigned: from 0 to 4,294,967,295, which equals $2^{32} - 1$

2. Number Systems
In general, number systems are sets of characters (symbols) and the rules for their use to represent numbers. The signs (symbols) used for displaying numbers are called figures. Number systems that have been used in the past can be categorized into two groups, *positional* and *non-positional*.

Positional number systems are those wherein the weight of a figure (its share in the total number of values) is determined based on its position in a number (a greater

MATHEMATICAL FUNDAMENTALS OF COMPUTER SCIENCE
Number Systems

---

position shows a bigger share of the value of the number).

Number systems, as we use them, consist of a basic set N of digits or letters, such as N = {0, 1, 2,...,9}, and a base B, for example B = 10.

In general, a number N of the form

$$N_B = a_{n-1} \, a_{n-2}...a_1 a_0$$

with word length n has the value

$$N_B = \sum_{i=0}^{n-1} a_i B^i = a_{n-1} B^{n-1} + a_{n-2} B^{n-1} + ... + a_1 B^1 + a_0 B^0$$

where

a_i = the figure in the number system $(0 \le a_i \le B-1)$
B = the base or radix of the number system $(B > 1)$
n = number of digits of a whole number.

In a *non-positional* number system, a number is a string whose value is equal to the sum of its component figures. For example, the ancient Romans used a non-positional number system.

MATHEMATICAL FUNDAMENTALS OF COMPUTER SCIENCE
Number Systems

The figures of the number system are I, V, X, L, C, D, and F, whose values are, respectively, 1, 5, 10, 50, 100, 500 and 1000. In the Roman numeral III, the digit I always has the same value of 1, and the value of the number is obtained by adding the values of all digits. In the Roman numerals IX and XI, the figure I again has the same value of 1.

These systems did not allow for mathematical operations such as are available with positional number systems. Non-positional number systems are not used in computers.

The following table shows number systems that are useful for computers.

NAME	BASE	DIGITS
Binary	2	0, 1
Octal	8	0, 1, 2, 3, 4, 5, 6, 7
Decimal	10	0, 1, 2, 3, 4, 5, 6, 7, 8, 9
Hexadecimal	16	0, 1, 2, 3, 4, 5, 6, 7, 8, 9, A, B, C, D, E, F

3. Decimal Number System

The number system that we know from everyday life is a decimal number system. In this number system, the base

MATHEMATICAL FUNDAMENTALS OF COMPUTER SCIENCE
Number Systems

$B = 10$. A decimal number represented by *abc.de* has a value of

$$N_{10} = a \cdot 10^2 + b \cdot 10^1 + c \cdot 10^0 + d \cdot 10^{-1} + e \cdot 10^{-2}$$

where the values of figures *a*, *b*, *c*, *d*, and *e* are taken from the group of digits $A = \{0, 1, 2, 3, 4, 5, 6, 7, 8, 9\}$.

Example:
The number 185.71 can be represented as follows:

$$185.71_{10} = 1 \cdot 10^2 + 8 \cdot 10^1 + 5 \cdot 10^0 + 7 \cdot 10^{-1} + 1 \cdot 10^{-2}$$
$$= 100 + 80 + 5 + 0.7 + 0.01$$

4. Binary Number System

A decimal number system proves to be very difficult to implement using electronic circuits for computing. For this purpose, a binary system is used. The binary number system has base $B = 2$, and numbers are recorded by figures from a set $A = \{0, 1\}$.

In the binary number system, the number *abc.de* has a value of:

$$N_2 = a \cdot 2^2 + b \cdot 2^1 + c \cdot 2^0 + d \cdot 2^{-1} + e \cdot 2^{-2}$$

MATHEMATICAL FUNDAMENTALS OF COMPUTER SCIENCE
Number Systems

---

where the values of figures a, b, c, d, and e are taken from the group of digits $A = \{0,1\}$.

Example:

The binary number 1011_2 represents the value

$$1011_2 = 1 \cdot 2^3 + 0 \cdot 2^2 + 1 \cdot 2^1 + 1 \cdot 2^0 = 8 + 2 + 1 = 11$$

In order to distinguish numbers written in many different systems, we add the value of the base of the number system as an index to the number.

Example:

$$1011_2 = 11_{10}$$

5. Decimal to Binary Number Conversion

The process of conversion from decimal to binary is different for integers and for the fractional part of the decimal representation of numbers.

a) An easy method of converting the integer part of a decimal number to its binary number equivalent is to write down the decimal number and go through the following steps:

MATHEMATICAL FUNDAMENTALS OF COMPUTER SCIENCE
Number Systems

Step 1: Divide the integer part of the decimal number to be converted by the value of the new base ($B = 2$).

Step 2: Use the remainder from Step 1 as the rightmost digit (least significant digit) of the new base number.

Step 3: Divide the quotient of the previous division by the new base.

Step 4: Record the remainder from Step 3 as the next digit (to the left) of the new base number.

Repeat Steps 3 and 4, getting remainders from right to left, until the quotient becomes zero in Step 3.

The last remainder thus obtained will be the most significant digit (MSD) of the new base number.

b) Converting the fractional part of the decimal of a number to its binary equivalent is also an interactive process.

Step 1: Multiply the fractional part of the decimal number by the base of the binary ($B = 2$), and write the integer part of the result obtained (0 or 1).

MATHEMATICAL FUNDAMENTALS OF
COMPUTER SCIENCE
Number Systems

Step 2 and following: Repeat Step 1, in each multiplication
using the fractional part obtained in the
previous multiplication. When the fractional
part becomes 0 (zero) or the desired accuracy
is achieved, the operation is terminated.

The integer parts, written in the order in which they were
obtained (from first to last), form the required binary part
of the fractional number.

Example:

Convert the decimal number 23.14_{10} into its binary

number equivalent.

Solution:

    a) Convert the integer part, 23, of the decimal number
       to binary:

Step 1:	$23 : 2 = 11$ remainder 1 (LSB)
Step 2:	$11 : 2 = 5$ " 1
Step 3:	$5 : 2 = 2$ " 1 ⇑
Step 4:	$2 : 2 = 1$ " 0
Step 5:	$1 : 2 = 0$ remainder 1 (MSB)

The binary result is obtained by placing all the
remainders in order from most significant bit (MSB) to
least significant bit (LSB).

The result is: $23_{10} = 10111_2$

  b) Converting the fractional part 0.14 of the decimal number to binary:

Step 1:	$0.14 \times 2 = 0.28$	0
Step 2:	$0.28 \times 2 = 0.56$	0
Step 3:	$0.56 \times 2 = 1.12$	1
Step 4:	$0.12 \times 2 = 0.24$	0 $\Downarrow$
Step 5:	$0.24 \times 2 = 0.48$	0
Step 6:	$0.48 \times 2 = 0.96$	0
Step 7:	$0.96 \times 2 = 1.92$	1
Step 8:	$0.92 \times 2 = 1.84$	1

When the desired accuracy is achieved, the operation is terminated.

The result is: $0.14_{10} = .00100011_2$

Therefore: $23.14_{10} = 10111.00100011_2$

6. Counting in Binary
Positive numbers can add leading zeros (0), and negative numbers leading ones (1), in front of the most significant digit (MSB) such that the value of the numbers does not change.

MATHEMATICAL FUNDAMENTALS OF COMPUTER SCIENCE
Number Systems

The binary counting sequence is shown in the following table.

Decimal	Binary	Decimal	Binary
0	00000	16	10000
1	00001	17	10001
2	00010	18	10010
3	00011	19	10011
4	00100	20	10100
5	00101	21	10101
6	00110	22	10110
7	00111	23	10111
8	01000	24	11000
9	01001	25	11001
10	01010	26	11010
11	01011	27	11011
12	01100	28	11100
13	01101	29	11101
14	01110	30	11110
15	01111	31	11111

7. Octal Number System

The octal number system has a base $B = 8 = 2^3$, and is recorded by the set of digits $A = \{0,1,2,3,4,5,6,7\}$.

MATHEMATICAL FUNDAMENTALS OF COMPUTER SCIENCE
Number Systems

In the octal number system, the number *abc.de* has the value

$$N_8 = a \cdot 8^2 + b \cdot 8^1 + c \cdot 8^0 + d \cdot 8^{-1} + e \cdot 8^{-2}$$

where the values of figures *a*, *b*, *c*, *d*, and *e* are taken from the group of digits $A = \{0,1,2,3,4,5,6,7\}$

Example:
Convert the octal number 107.1 to the decimal representation form.

Solution:

$$107.1_8 = 1 \cdot 8^2 + 0 \cdot 8^1 + 7 \cdot 8^0 + 1 \cdot 8^1$$

$$= 64 + 0 + 7 + 0.125 = 71.125_{10}$$

Hence:

$$107.1_8 = 71.125_{10}$$

8. Decimal to Octal Number Conversion

This type of conversion is executed in an analogous manner to the decimal to binary number conversion: dividing the integer part, or multiplying the fractional part, of the decimal number by the counting base of the octal number system ($B = 8$). The last remainder thus obtained by dividing will be the most significant digit (MSD) of the octal base number.

MATHEMATICAL FUNDAMENTALS OF COMPUTER SCIENCE
Number Systems

For the fractional part, write the integer parts of the products in the order in which they are calculated.

Example:
Record the decimal number 245_{10} in the octal representation form.

Solution:

$$245 : 8 = 30 \text{ remainder } 5 \text{ (LSB)}$$
$$30 : 8 = 3 \quad " \quad 6 \Uparrow$$
$$3 : 8 = 0 \quad " \quad 3 \text{ (MSB)}$$

Accordingly, $245_{10} = 365_8$

9. Binary to Octal Number Conversion and Its Converse

To go from a binary to an octal number system, it is enough to write the number in groups of three binary digits and wrie each of them in decimal

Example 1:
Write down the binary number 1010100110.10 in octal form.

Solution:
- Split the number into groups of three binary digits.

MATHEMATICAL FUNDAMENTALS OF COMPUTER SCIENCE
Number Systems

- Add two zeros to the left of the integer part of the observed number to make a group of three on the far left.
- Add a zero to the right binary digits.
- Interpret in the decimal system the individually grouped binary digits.

$$001 \ 010 \ 100 \ 110 \ . \ 100$$
$$1 \quad 2 \quad 4 \quad 6 \quad 4$$

Hence,

$$1010100110.10_2 = 1246.4_8$$

Example 2:

Find the binary equivalent of the octal number 6125.3.

Solution:

$$6 \quad 1 \quad 2 \quad 5 \ . \ 3$$
$$110 \ 001 \ 010 \ 101 \ \ 011$$

The required binary number is 110001010101.011_2

10. Counting in Octal

The octal counting sequence is shown in the following table.

Decimal	Octal	Decimal	Octal
0	00	16	20
1	01	17	21

MATHEMATICAL FUNDAMENTALS OF COMPUTER SCIENCE
Number Systems

Continued from # 10

2	02	18	22
3	03	19	23
4	04	20	24
5	05	21	25
6	06	22	26
7	07	23	27
8	10	24	30
9	11	25	31
10	12	26	32
11	13	27	33
12	14	28	34
13	15	29	35
14	16	30	36
15	17	31	37

11. Hexadecimal Number System
The hexadecimal number system has the base
$B = 16 = 2^4$. To record numbers in this system
it is necessary to have 16 different digits available.

The set of digits of the hexadecimal number system is

$$A = \{0, 1, 2, 3, 4, 5, 6, 7, 8, 9, A, B, C, D, E, F\}.$$

MATHEMATICAL FUNDAMENTALS OF COMPUTER SCIENCE
Number Systems

In the hexadecimal number system, the number *abc.de* has a value of

$$N_{16} = a \cdot 16^2 + b \cdot 16^1 + c \cdot 16^0 + d \cdot 16^{-1} + e \cdot 16^{-2}$$

where the values of figures *a*, *b*, *c*, *d*, and *e* are taken from the group of digits

$$A = \{0, 1, 2, 3, 4, 5, 6, 7, 8, 9, A, B, C, D, E, F\}$$

Example:
Convert the hexadecimal number $FB40A.2_{16}$ to decimal form.

Solution:
$$FB40A.2_{16} = F \cdot 16^4 + B \cdot 16^3 + 4 \cdot 16^2 + 0 \cdot 16^1 + A \cdot 16^0 + 2 \cdot 16^{-1}$$
$$= 15 \cdot 16^4 + 11 \cdot 16^3 + 4 \cdot 16^2 + 0 \cdot 16^1 + 10 \cdot 16^0 + 2 \cdot 16^{-1}$$
$$= 1029130.125_{10}$$

12. Binary to Hexadecimal Number Conversion and Its Converse

A binary number is converted to hexadecimal form by grouping every four binary figures and then interpreting them in individual hexadecimal shape.

MATHEMATICAL FUNDAMENTALS OF COMPUTER SCIENCE
Number Systems

Example 1:

Write the binary number 100000111101010_2 in hexadecimal form.

Solution:

$$\textbf{0}100 \ 1001 \ 1110 \ 1010$$
$$4 \quad 9 \quad E \quad A$$

Hence,

$$100100111101010_2 = 49EA_{16}$$

Example 2:

Show the hexadecimal number $A87_{16}$ in binary form.

Solution:

$$A \quad 8 \quad 7$$
$$1010 \quad 1000 \quad 0111$$

Therefore,

$$A87_{16} = 101010000111_2$$

13. Counting in Hexadecimal

The hexadecimal counting sequence is shown in the following table.

MATHEMATICAL FUNDAMENTALS OF COMPUTER SCIENCE
Number Systems

Decimal	Hexadecimal	Decimal	Hexadecimal
0	00	16	10
1	01	17	11
2	02	18	12
3	03	19	13
4	04	20	14
5	05	21	15
6	06	22	16
7	07	23	17
8	08	24	18
9	09	25	19
10	0A	26	1A
11	0B	27	1B
12	0C	28	1C
13	0D	29	1D
14	0E	30	1E
15	0F	31	1F

14. Comparison of Number Systems

A comparison of the different number systems' counting sequences is shown in the following table.

Decimal	Binary	Octal	Hexadecimal
0	00000	00	00
1	00001	01	01
2	00010	02	02

MATHEMATICAL FUNDAMENTALS OF COMPUTER SCIENCE
Number Systems

Continued from # 14

3	00011	03	03
4	00100	04	04
5	00101	05	05
6	00110	06	07
7	00111	07	07
8	01000	10	08
9	01001	11	09
10	01010	12	0A
11	01011	13	0B
12	01100	14	0C
13	01101	15	0D
14	01110	16	0E
15	01111	17	0F
16	10000	20	10
17	10001	21	11
18	10010	22	12
19	10011	23	13
20	10100	24	14
21	10101	25	15
22	10110	26	16
23	10111	27	17
24	11000	30	18
25	11001	31	19
26	11010	32	1A

MATHEMATICAL FUNDAMENTALS OF COMPUTER SCIENCE
Binary Logic

Continued from # 14

27	11011	33	1B
28	11100	34	1C
29	11101	35	1D
30	11110	36	1E
31	11111	37	1F

15. Binary Logic

Binary logic means that the logical operations AND, OR and NOT are defined with designated variables, such as x, y, and z. These variables can have only two values, 0 and 1.

a) The definition of the logical operation NOT is shown in the following table.

Logical operation NOT	
x	$\bar{x} = z$
0	1
1	0

b) The definitions of the logical operations AND and OR are shown in the following table.

MATHEMATICAL FUNDAMENTALS OF COMPUTER SCIENCE
Binary Arithmetic

Logical operations AND & OR			
x	y	$x \cdot y = z$	$x + y = z$
0	0	0	0
0	1	0	1
1	0	0	1
1	1	1	1

16. Binary Arithmetic

Binary arithmetic is a branch of mathematics that studies operations with numbers that follow the rules of Boolean algebra. Each operand end result takes one of two values, 0 or 1.

For any number system, even for binary, it is necessary to define its basic operations (addition, subtraction, multiplication and division):

a) Binary addition

Basic arithmetic operations in the binary number system are performed by defined rules, so the addition of binary digits can be performed in four ways according to the following:

$$0 + 0 = 0 \text{ transfer } 0$$
$$0 + 1 = 1 \text{ transfer } 0$$
$$1 + 0 = 1 \text{ transfer } 0$$
$$1 + 1 = 0 \text{ transfer } 1$$

MATHEMATICAL FUNDAMENTALS OF COMPUTER SCIENCE
Binary Arithmetic

"Transmission" refers to the transfer that occurs from the current to the next (more) weighted location of the observed binary number.

Example:

Add the binary numbers 11011_2 and 1011_2.

Solution:

Transfer	1	1		1	1	
		1	1	0	1	1
+			1	0	1	1
Summation	1	0	0	1	1	0

b) Binary subtraction

The subtraction of numbers can be reduced to addition. To make this possible, the subtrahend must be converted into a negative number. Negative numbers in the binary number system are represented using the two's complement.

Two's complement:

1. The minuend and the subtrahend are written with the same number of digits (add to the left of the subtrahend the required number of zeros).

MATHEMATICAL FUNDAMENTALS OF COMPUTER SCIENCE
Binary Arithmetic

---

2. Each 0 in the subtrahend is converted to 1 and each 1 is converted to 0; the resulting number is called the complement number.

3. The complement is added to 1, generating the two's complement.

Example 1:
Define the two's complement of a binary subtrahend number.

Solution:

Minuend		1	1	0	1	1
Subtrahend		**0**	1	0	1	1
Complement		1	0	1	0	0
	+					1
Two's complement		1	0	1	0	1

Example 2:
Subtract binary numbers 11011_2 and 1011_2

Solution:
The operation of subtraction of the numbers 11011_2 and 1011_2 is shown in the following table:

MATHEMATICAL FUNDAMENTALS OF COMPUTER SCIENCE
Binary Arithmetic

Transfer		1	1	1	1	1	
Minuend			1	1	0	1	1
Two's complement	+		1	0	1	0	1
Difference		1	1	0	0	0	0

After the minuend is added to the two's complement of subtrahend, the result is not correct. To get the correct result, the far left digit 1 should be discarded, so the result is 10000_2

Binary subtraction can be also performed using the rule according to the following:

$$0 - 0 = 0 \text{ transfer } 0$$
$$0 - 1 = 1 \text{ transfer } 1$$
$$1 - 0 = 1 \text{ transfer } 0$$
$$1 - 1 = 0 \text{ transfer } 0$$

Example:

Subtract the two binary numbers 1101001_2 and 10011_2

Solution:

```
    1 1 0 1 0 0 1
  -     1 0 0 1 1
Transfer   1   1 1
  _____
    1 0 1 0 1 1 0
```

MATHEMATICAL FUNDAMENTALS OF COMPUTER SCIENCE
Binary Arithmetic

---

c) Binary multiplication

For binary multiplication, the same procedure as for multiplication in the decimal number system is applied, but it is essentially simpler, since only ones and zeros are used.

Example:

Multiply the binary numbers 1011_2 and 101_2

Solution:

$$
\begin{array}{ll}
\text{Multiplicand} & 1011 \\
\text{Multiplier} & \times\ 101 \\
\hline
& 1011 \\
& 0000 \\
& 1011 \\
\hline
\text{Result} & 110111 \\
\end{array}
$$

d) Binary division

The procedure of division in the binary number system is identical to that in the decimal number system.

Example:

Binary divide the number 53 by the number 7 given in decimal number system.

MATHEMATICAL FUNDAMENTALS OF COMPUTER SCIENCE
Representation of Signed Numbers for Computers

Solution:

Dividing binary numbers

$$53_{10} = 110101_2 \text{ and } 7_{10} = 111_2$$

Dividend   Divisor   Quotient

$\quad$ 110101 : 111 $\quad$ = $\quad$ 0111

$-$ 000

$\quad$ 1101

$-$ 111

$\quad$ 1100

$-$ 111

$\quad$ 1011

$-$ 111

$\quad\quad$ 100 Remainder

17. Representation of Signed Numbers in Computers

There are two ways to represent signed whole binary numbers:

1. Sign and absolute value
2. Two's complement

MATHEMATICAL FUNDAMENTALS OF COMPUTER SCIENCE
Representation of Signed Numbers for Computers

---

a) Sign and absolute value

To the absolute value of the binary number in front of the number of highest weight is added the number 0 if the number is positive, or 1 if the number is negative.

Example:

$$5_{10} = 101_2 \quad \text{unmarked binary number}$$

$$+5_{10} = 0101 \quad \text{marked positive binary number}$$

$$-5_{10} = 1101 \quad \text{marked negative binary number}$$

b) Two's complement

1. A positive integer is obtained by adding the digit 0 in front of the binary entry corresponding to the absolute value of the number.

2. A negative integer can be obtained by the following steps:

Step 1: In front of the representation of the absolute value of the number, add the digit 0.

Step 2: Invert all binary digits (i.e., 0 converts to 1; 1 converts to 0).

Step 3: The resulting number is added to 1.

MATHEMATICAL FUNDAMENTALS OF COMPUTER SCIENCE
Representation of Signed Numbers for Computers

Example:

Represent the number -7_{10} as a two's complement.

Solution:

$$7_{10} = 111_2$$

add zero in front

$$+ 7_{10} = 0111_2$$

invert digits of positive number

$$1000 \quad +$$
$$1$$

$$-7_{10} = 1001_2$$

A simplified procedure for recording a binary number as a two's complement:

Step 1: The initial binary number is split into two parts, left and right, so that the right part consists of the first digit 1 from the right, and all the zeros following the digit 1, while the remaining digits make up the left part of the number.

MATHEMATICAL FUNDAMENTALS OF COMPUTER SCIENCE
Representation of Signed Numbers for Computers

Step 2: The two's complement is obtained by inverting all digits on the left side of the number (ones are replaced with zeros, and zeros are replaced with ones) while leaving the right part unchanged.

Example: Calculate the two's complement of the binary number 01010010010000_2

Solution:
Split the number into two parts:
010100100 | 10000
left side   right side
invert the digits on the left side only
Two's complement = 10101101110000

Attributing the two's complement: positive and negative numbers can be added by putting leading zeros or leading ones in front of the most significant digit, which leaves the values of the numbers

Example:
Positive number: $0111_2 = 0000000111_2$
Negative number: $1001_2 = 11111001_2$

Determining decimal values: the decimal value of the binary number X written as a two's complement of

$n + 1$ numbers is found by applying the following formula:

$$X = -a_n 2^n + a_{n-1} 2^{n-1} + ... + a_1 2^1 + a_0 2^0$$

Example:
Determine the decimal values of these numbers written as two's complements: 0101 and 1011

Solution:
$$0101_2 = -0 \cdot 2^3 + 1 \cdot 2^2 + 0 \cdot 2^1 + 1 \cdot 2^0 = 4 + 1 = 5$$
$$1011_2 = -1 \cdot 2^3 + 0 \cdot 2^2 + 1 \cdot 2^1 + 1 \cdot 2^0 = -8 + 2 + 1 = -5$$

18. Range of Numbers Written with n Digits
a) The range of unsigned integers written with n digits in binary form is obtained by the following formula:
$$0 \le x \le 2^n - 1 \implies x \in \left\{ 0, 1, ..., 2^n - 1 \right\}$$

The range of integers that can be represented by 4 and 8 binary figures is:

$$n = 4 \implies 0 \le x \le 15 \implies x \in \left\{ 0, 1, ..., 15 \right\}$$
$$n = 8 \implies 0 \le x \le 255 \implies x \in \left\{ 0, 1, ..., 255 \right\}$$

MATHEMATICAL FUNDAMENTALS OF
COMPUTER SCIENCE
Representation of Signed Numbers for Computers

---

b) The range of signed integers written as two's complements with n digits is obtained by the following formula:

$$x \le 2^{n-1} - 1 \quad \text{for} \ x \ge 0,$$

$$|x| \le 2^{n-1} \ \text{for} \ x < 0 \ \Rightarrow x \in \left\{-2^{n-1},...,-1,0,1,...,2^{n-1}-1\right\}$$

The range of numbers that can be represented by 4 and 8 binary figures is:

$$n = 4 \ \Rightarrow -8 \le x \le 7 \quad \Rightarrow x \in \left\{-8,...,-1, 0, 1,...,7\right\}$$

$$n = 8 \ \Rightarrow -128 \le x \le 127 \Rightarrow x \in \left\{-128,...,-1,0,1,...,127\right\}$$

19. Representation of Real Numbers

For the representation of real numbers (numbers with decimal points), the *floating point* format is used.
A number recorded in floating point has three components: the sign, the exponent, and the mantissa.

31   **0**

Sign	Exponent	Mantissa

The decimal value of a number recorded in floating point is obtained by the following formula:

$$(\text{Sign}) \cdot \text{Mantissa} \cdot 2^{\text{Exponent}}$$

MATHEMATICAL FUNDAMENTALS OF COMPUTER SCIENCE
Representation of Signed Numbers for Computers

There are various standards that define how many bits are used for each component as well as what format the components are written in.

The generally accepted standard notation for floating point is IEEE Standard 754. To record a number using the floating point according to IEEE Standard 754, use

1 bit for the sign
8 bits for the exponent
23 bits for the mantissa.

20. Representation of Character Data Types

The character set consists of:

- upper- and lowercase letters of the alphabet,
- decimal digits,
- special signs (signs on the keyboard that are not letters or numbers and can be printed, for example: !, #, $, %, and so on),
- control means (cannot be printed or displayed on a screen used to manage input/output device; examples are sound signals and the like).

There are several methods for the binary representation of characters on a computer. The best known of these is

MATHEMATICAL FUNDAMENTALS OF COMPUTER SCIENCE
Information in Binary Form

ASCII, the American Standard Code for Information Interchange.

21. Information in Binary Form

1 bit [1b]: information from one binary digit

1 byte [1B]: 8 bits of information

1 kilobyte [1KB]: 1024 bytes of information

1 megabyte [1MB]: 1024 kilobytes or
1024 x 1024 = 1,048,576
bytes of information

1 gigabyte [1GB]: 1024 megabytes or
1024 x1024 x 1024 =1,073,741,824
bytes of information.

PART **III**

PHYSICS

Physics is the science of nature in the broadest sense. Physicists study the behavior and properties of matter in a wide variety of contexts and scales, ranging from the sub-microscopic particles from which all ordinary matter is made (particle physics) to the material universe as a whole (cosmology).

This part of the book contains the most frequently used formulas and definitions related to the following:

- Mechanics
- Mechanics of Fluids
- Temperature and Heat
- Electricity and Magnetism
- Light
- Wave Motion and Sound

MECHANICS

In physics, classical mechanics is one of the two major sub-fields in the science of mechanics (quantum mechanics is the other). Classical mechanics is concerned with the motions of bodies and the forces that cause those motions. This subject concerns macroscopic bodies, that is, bodies that can be easily seen in the solid state.

This section contains the most frequently used formulas, rules, and definitions related to the following:

- Kinematics
- Dynamics
- Statics

MECHANICS
Kinematics

---

1. Scalars and Vectors

The mathematical quantities that are used to describe the motion of objects can be divided into two categories: scalars and vectors.

a) Scalars
Scalars are quantities that can be fully described by a magnitude alone.

b) Vectors
Vectors are quantities that can be fully described by both a magnitude and a direction.

2. Distance and Displacement

a) Distance
Distance is a scalar quantity that refers to how far an object has gone during its motion.

b) Displacement
Displacement is the change in position of a object. It is a vector that includes the magnitude as a distance, such as five miles, and a direction, such as north.

3. Acceleration
Acceleration is the change in velocity per unit of time. Acceleration is a vector quantity.

MECHANICS
Kinematics

---

4. Speed and Velocity

a) Speed
The distance traveled per unit of time is called the speed, for example 35 miles per hour. Speed is a scalar quantity.

b) Velocity
The quantity that combines both the speed of an object and its direction of motion is called velocity.
Velocity is a vector quantity.

5. Frequency
Frequency is the number of complete vibrations per unit time in simple harmonic or sinusoidal motion.

6. Period
Period is the time required for one full cycle of vibration or oscillation. It is the reciprocal of the frequency.

7. Angular Displacement
Angular displacement is the rotational angle through which any point on a rotating body moves.

8. Angular Velocity
Angular velocity is the angular displacement per unit of time.

MECHANICS
Kinematics

9. Angular Acceleration
Angular acceleration is the rate of change of angular velocity with respect to time.

10. Rotational Speed
Rotational speed is the number of revolutions (a revolution is one complete rotation of a body) per unit of time.

11. Uniform Linear Motion
A path is a straight line. The total distance traveled corresponds with the rectangular area in the graph of v versus t.

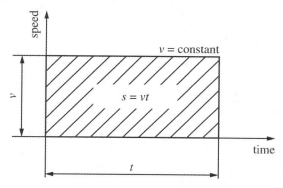

a) Distance: $s = vt$

b) Speed: $v = \dfrac{s}{t}$

where

       s = distance (m)
       v = speed (m/s)
       t = time (s)

12. Uniform Accelerated Linear Motion

1) If $v_0 > 0$ and $a > 0$, then

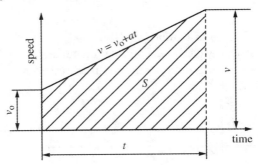

a) Distance: $s = v_0 t + \dfrac{at^2}{2}$

b) Speed: $v = v_0 + at$

where

       s = distance (m)
       v = speed (m/s)
       t = time (s)
       v_0 = initial speed (m/s)

       a = constant acceleration (m/s^2)

MECHANICS
Kinematics

---

2) If $v_0 = 0$ and $a > 0$, then

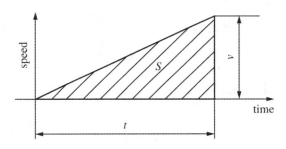

a) Distance: $$s = \frac{at^2}{2}$$

The shaded area in the *v–t diagram* represent the distance *s* traveled during the time period *t*.

b) Speed:
$$v = a \cdot t$$
where
s = distance (m)
v = speed (m/s)
v_0 = initial speed (m/s)
a = constant acceleration (m/s^2)

13. Rotational Motion
Rotational motion occurs when the body itself is spinning. The path is a circle about the axis of rotation.

MECHANICS
Kinematics

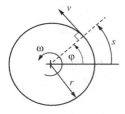 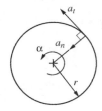

a) Distance:
$$s = r\varphi$$

b) Velocity:
$$v = r\omega$$

c) Tangential acceleration:
$$a_t = r \cdot \alpha$$

d) Centripetal acceleration:
$$a_n = \omega^2 r = \frac{v^2}{r}$$

where

$\widehat{\varphi}$ = angle determined by s and r (rad)

ω = angular velocity $\left(s^{-1}\right)$

α = angular acceleration $\left(1/s^2\right)$

a_t = tangential acceleration $\left(1/s^2\right)$

a_n = centripetal acceleration $\left(1/s^2\right)$

The distance s, velocity v, and tangential acceleration a_t are proportional to the radius r.

MECHANICS
Kinematics

---

14. Uniform Rotation about a Fixed Axis

ω_0 = constant; $\alpha = 0$

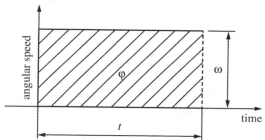

a) Angle of rotation:

$$\varphi = \omega \cdot t$$

b) Angular velocity:

$$\omega = \frac{\varphi}{t}$$

where

φ = angle of rotation (rad)

ω = angular velocity $\left(s^{-1}\right)$

α = angular acceleration $\left(1/s^2\right)$

ω_0 = initial angular speed $\left(s^{-1}\right)$

The shaded area in the $\omega - t$ diagram represents the angle of rotation φ covered during time period t.

MECHANICS
Kinematics

---

15. Uniform Accelerated Rotation about a Fixed Axis

1) If $\omega_0 > 0$ and $\alpha > 0$, then

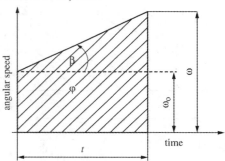

a) Angle of rotation:

$$\varphi = \frac{t}{2}(\omega_0 + \omega) = \omega_0 t + \frac{1}{2}\alpha t^2$$

b) Angular velocity:

$$\omega = \omega_0 + \alpha t = \sqrt{\omega_0^2 + 2\alpha\varphi}$$

$$\omega_0 = \omega - \alpha t = \sqrt{\omega^2 - 2\alpha\varphi}$$

c) Angular acceleration:

$$\alpha = \frac{\omega - \omega_0}{t} = \frac{\omega^2 - \omega_0^2}{2\varphi}$$

d) Time: $t = \dfrac{\omega - \omega_0}{\alpha} = \dfrac{2\varphi}{\omega_0 + \omega}$

2) If $\omega_0 = 0$ and α = constant, then

MECHANICS
Kinematics

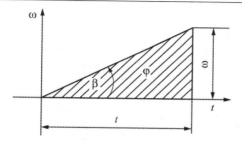

a) Angle of rotation:

$$\phi = \frac{\omega \cdot t}{2} = \frac{\alpha \cdot t^2}{2} = \frac{\omega^2}{2\alpha}$$

b) Angular velocity:

$$\omega = \sqrt{2\alpha\varphi} = \frac{2\varphi}{t} = \alpha \cdot t \; ; \; \omega_0 = 0$$

c) Angular acceleration:

$$\alpha = \frac{\omega}{t} = \frac{2\varphi}{t^2} = \frac{\omega^2}{2\varphi}$$

d) Time:

$$t = \sqrt{\frac{2\varphi}{\alpha}} = \frac{\omega}{\alpha} = \frac{2\varphi}{\omega}$$

16. Simple Harmonic Motion

Simple harmonic motion occurs when an object moves repeatedly over the same path in equal time intervals. The maximum deflection from the position of rest is called the "amplitude."

MECHANICS
Kinematics

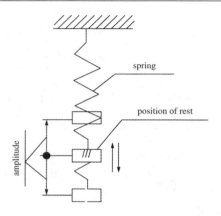

A mass vibrating on a spring is an example of an object in simple harmonic motion. The motion is sinusoidal in time and exhibits a single frequency.

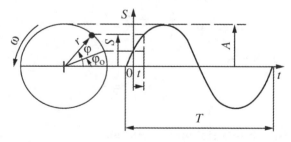

a) Displacement:
$$s = A\sin(\omega \cdot t + \varphi_0)$$

b) Velocity:
$$v = A\omega\cos(\omega \cdot t + \varphi_0)$$

c) Angular acceleration:

$$\alpha = -A\omega^2 \sin(\omega \cdot t + \varphi_0)$$

where

s = displacement

A = amplitude

φ_0 = angular position at time $t = 0$

φ = angular position at time t

T = period

17. Pendulum

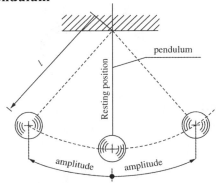

A pendulum consists of an object suspended so that it swings freely back and forth about a pivot.

a) Period:

$$T = 2\pi\sqrt{\frac{l}{g}}$$

where

   T = period (s)
   l = length of pendulum (m)
   $g = 9.81$ (m/s^2) or 32.2 (ft/s^2)

18. Free Fall

A free-falling object is an object that is falling due to the sole influence of gravity.

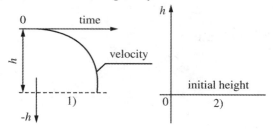

a) Initial speed:

$$v_0 = 0$$

b) Distance:

$$h = -\frac{gt^2}{2} = -\frac{vt}{2} = -\frac{v^2}{2g}$$

c) Speed:

$$v = +gt = -\frac{2h}{t} = \sqrt{-2gh}$$

d) Time:

$$t = +\frac{v}{g} = -\frac{2h}{v} = \sqrt{-\frac{2h}{g}}$$

19. Vertical Projection

a) Initial speed:

$$v_0 > 0 \text{ (upwards)}; \quad v_0 < 0 \text{ (downwards)}$$

b) Distance:

$$h = v_0 t - \frac{gt^2}{2} = (v_0 + v)\frac{t}{2}; \quad h_{max} = \frac{v_0^2}{2g}$$

c) Time:

$$t = \frac{v_0 - v}{g} = \frac{2h}{v_0 + v}; \quad t_{h\,max} = \frac{v_0}{g}$$

where
   v = velocity (m/s)
   h = distance (m)
   g = acceleration due to gravity (m/s^2)

20. Angled Projection
   Upwards $(\alpha > 0)$; downwards $(\alpha < 0)$

MECHANICS
Kinematics

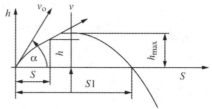

a) Horizontal distance:
$$s = v_0 \cdot t \cos \alpha$$

b) Altitude:
$$h = v_0 t \sin \alpha - \frac{g \cdot t^2}{2} = s \tan \alpha - \frac{g \cdot s^2}{2v_0^2 \cos \alpha}$$

$$h_{max} = \frac{v_0^2 \sin^2 \alpha}{2g}$$

c) Velocity:
$$v = \sqrt{v_0^2 - 2gh} = \sqrt{v_0^2 + g^2 t^2 - 2gv_0 t \sin \alpha}$$

d) Time:
$$t_{h_{max}} = \frac{v_0 \sin \alpha}{g}; \; t_{s1} = \frac{2v_0 \sin \alpha}{g}$$

21. Horizontal Projection $(\alpha = 0)$

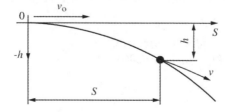

a) Horizonal distance:

$$s = v_0 t = v_0 \sqrt{\frac{2h}{g}}$$

b) Altitude:

$$h = -\frac{gt^2}{2}$$

c) Trajectory velocity:

$$v = \sqrt{v_0^2 + g^2 t^2}$$

where

v_0 = initial velocity (m/s)
v = trajectory velocity (m/s)
s = horizontal distance (m)
h = height (m)

22. Sliding Motion on an Inclined Plane

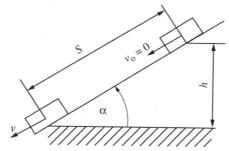

MECHANICS
Kinematics

---

1) If excluding friction ($\mu = 0$), then

    a) Velocity:

$$v = at = \frac{2s}{t} = \sqrt{2as}$$

    b) Distance:

$$s = \frac{at^2}{2} = \frac{vt}{2} = \frac{v^2}{2a}$$

    c) Acceleration:

$$a = g \sin \alpha$$

2) If including friction ($\mu > 0$), then

    a) Velocity:

$$v = at = \frac{2s}{t} = \sqrt{2as}$$

    b) Distance:

$$s = \frac{at^2}{2} = \frac{vt}{2} = \frac{v^2}{2a}$$

    c) Acceleration:

$$a = g (\sin \alpha - \mu \cos \alpha)$$

where
  μ = coefficient of sliding friction
  g = acceleration due to gravity,
        9.81 (m/s²)
  v_0 = initial velocity (m/s)
  v = trajectory velocity (m/s)
  s = distance (m)
  a = acceleration (m/s²)
  α = inclined angle

23. Rolling Motion on an Inclined Plane

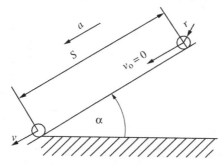

1) If excluding friction $(f = 0)$, then

a) Velocity:

$$v = at = \frac{2s}{t} = \sqrt{2as}$$

b) Acceleration:

$$a = \frac{gr^2}{r^2 + k^2} \sin \alpha$$

c) Distance:

$$s = \frac{at^2}{2} = \frac{vt}{2} = \frac{v^2}{2a}$$

d) Tilting angle:

$$\tan \alpha = \mu_0 \frac{r^2 + k^2}{k^2}$$

2) If including friction $(f > 0)$, then
   a) Distance:

$$s = \frac{at^2}{2} = \frac{vt}{2} = \frac{v^2}{2a}$$

b) Velocity:

$$v = at = \frac{2s}{t} = \sqrt{2as}$$

c) Acceleration:

$$a = gr^2 \frac{\sin \alpha - (f/r)\cos \alpha}{r^2 + k^2}$$

d) Tilting angle:

$$\tan \alpha_{min} = \frac{f}{r}; \quad \tan \alpha_{max} = \mu_0 \frac{r^2 + k^2 - fr}{k^2}$$

The value of k can be calculated by the formulas given in the following table.

Ball	Solid cylinder	Pipe with low wall thickness
$k^2 = \dfrac{2r^2}{5}$	$k^2 = \dfrac{r^2}{2}$	$k^2 = \dfrac{r_i^2 + r_o^2}{2} \approx r^2$

where

s = distance (m)

v = velocity (m/s)

a = acceleration (m/s^2)

α = tilting angle ($^\circ$)

f = lever arm of rolling resistance (m)

k = radius of gyration (m)

μ_0 = coefficient of static friction

g = acceleration due to gravity (m/s^2)

24. Newton's First Law of Motion
Newton's First Law or the Law of Inertia:
An object that is in motion continues in motion with the same velocity at constant speed and in a straight line, and an object at rest continues at rest unless an unbalanced (outside) force acts upon it.

MECHANICS
Dynamics

---

25. Newton's Second Law
The Second Law of Motion, called the Law of Acceleration, states that:
The total force acting on an object equals the mass of the object times its acceleration.
In equation form, this law is

$$F = ma$$

where

F = total force acting on an object (N)

m = mass of object (kg)

a = acceleration (m/s^2)

26. Newton's Third Law
The Third Law of Motion, called the Law of Action and Reaction, can be stated as follows:
For every force applied by object A to object B (action), there is a force exerted by object B on object A (the reaction) which has the same magnitude but is opposite in direction.
In equation form this law is

$$F_B = -F_A$$

where

F_B = force of action (N)

F_A = force of reaction (N)

MECHANICS
Dynamics

27. Momentum of Force

Momentum can be defined as mass in motion.
Momentum is a vector quantity; in other words, the
direction is important:

$$p = mv$$

28. Impulse of Force

The impulse of a force is equal to the change in
momentum that the force causes in an object:

$$I = Ft$$

where

p = momentum (N s)

m = mass of object (kg)

v = velocity of object (m/s)

I = impulse of force (N s)

F = force (N)

t = time for which force is applied (s)

29. Law of Conservation of Momentum

One of the most powerful laws in physics is the law of
momentum conservation, which can be stated as follows:
*In the absence of external forces, the total momentum of
the system is constant.* For example,

MECHANICS
Dynamics

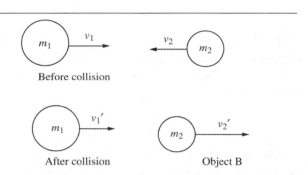

Before collision

After collision   Object B

If two objects of mass m_1 and mass m_2, having velocities v_1 and v_2, collide and then separate with velocities v_1' and v_2', the equation for the conservation of momentum is

$$m_1 v_1 + m_2 v_2 = m_1 v_1' + m_2 v_2'$$

30. Friction

Friction is a force that always acts parallel to the surfaces in contact and opposite to the direction of motion. Starting friction is greater than moving friction. Friction increases as the force between the surfaces increas

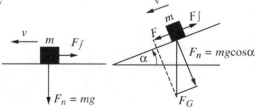

The characteristics of friction can be described by the following equation:

MECHANICS
Dynamics

$$F_f = \mu F_n$$

where

F_f = frictional force (N)

F_n = normal force (N)

μ = coefficient of friction ($\mu = \tan\alpha$)

31. General Law of Gravity

Gravity is a force that attracts bodies of matter toward each other. Simply put, gravity is the attraction between any two objects that have mass.

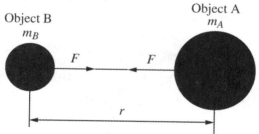

The general formula for gravity is

$$F = \Gamma \frac{m_A m_B}{r^2}$$

where

m_A, m_B = mass of objects A and B (kg)

F = magnitude of attractive force between objects A and B (N)

r = distance between objects A and B (m)

Γ = gravitational constant ($m^3 kg^{-1} s^{-2}$)

MECHANICS
Dynamics

---

32. Gravitational Force
The force of gravity is given by the equation

$$F_G = g \frac{R_e^2 m}{(R_e + h)^2}$$

On the earth's surface, $h = 0$; so,

$$F_G = mg$$

where

F_G = force of gravity (N)

R_e = radius of the Earth ($R_e = 6.37 \times 10^6$ m)

m = mass (kg)

g = acceleration due to gravity (m/s^2)

$g = 9.81$ (m/s^2) or $g = 32.2$ (ft/s^2)

The acceleration of a falling body is independent of the mass of the object.

The weight F_w of an object is actually the force of gravity on that object:

$$F_W = mg$$

33. Centrifugal Force
Centrifugal force is the apparent force drawing a rotating body away from the center of rotation, and it is caused

---

by the inertia of the body. Centrifugal force can be
calculated by the formula

$$F_c = \frac{mv^2}{r} = m\omega^2 r$$

34. Centripetal Force

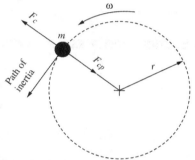

Centripetal force is defined as the force acting on a body
in curvilinear motion that is directed toward the center of
curvature or axis of rotation. Centripetal force is equal in
magnitude to centrifugal force but in the opposite
direction.

$$F_{cp} = -F_c = -\frac{mv^2}{r}$$

where

F_c = centrifugal force (N)
F_{cp} = centripetal force (N)
m = mass of the body (kg)
v = velocity of the body (m/s)

MECHANICS
Dynamics

---

r = radius of curvature of the
    path of the body (m)
ω = angular velocity $\left(s^{-1}\right)$

35. Torque

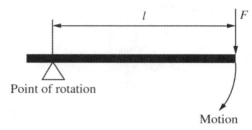

Torque is the ability of a force to cause a body to rotate about a particular axis.

Torque can have either a clockwise or a counterclockwise direction. To distinguish between the two possible directions of rotation, we adopt the convention that a counterclockwise torque is positive and a clockwise torque is negative.

One way to quantify torque is by

$$T = F \cdot l$$

where

    T = torque (N m or lb ft)
    F = applied force (N or lb)
    l = length of torque arm (m or ft)

MECHANICS
Dynamics

36. Work

Work is the product of a force in the direction of the motion and the displacement.

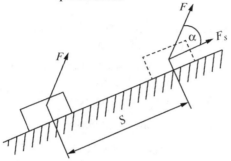

a) Work done by a constant force:

$$W = F_s \cdot s = F \cdot s \cdot \cos\alpha$$

where

W = work (N m = J)
F_s = component of force along the direction of movement (N)
s = distance the object is displaced (m)

b) Work done by a variable force

If the force is not constant along the path of the object, we need to calculate the force over very tiny intervals and then add them up. This is exactly what integration over differential small intervals of a line can accomplish:

MECHANICS
Dynamics

$$W = \int_{s_i}^{s_f} F_s(s) \cdot ds = \int_{s_i}^{s_f} F(s)\cos\alpha \cdot ds$$

where

$F_s(s)$ = component of the force function along the direction of movement (N)

$F(s)$ = function of the magnitude of the force vector along the displacement curve (N)

s_i = initial location of the body (m)

s_f = final location of the body (m)

α = angle between the displacement and the force

37. Energy

Energy is defined as the ability to do work. The quantitative relationship between work and mechanical energy is expressed by the equation

$$TME_i + W_{ext} = TME_f$$

where

TME_i = initial amount of total mechanical energy (J)

W_{ext} = work done by external forces (J)

TME_f = final amount of total mechanical energy (J)

MECHANICS
Dynamics

There are two kinds of mechanical energy: kinetic and potential.

a) Kinetic energy

Kinetic energy is the energy of motion. The following equation is used to represent the kinetic energy of an object:

$$E_k = \frac{1}{2}mv^2$$

where

m = mass of moving object (kg)
v = velocity of moving object (m/s)

b) Potential energy

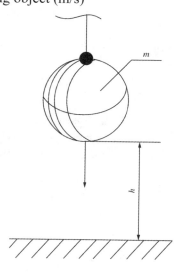

Potential energy is the stored energy of a body and is due to its internal characteristics or its position. Gravitational potential energy is defined by the formula

$$E_{pg} = m \cdot g \cdot h$$

MECHANICS
Dynamics

where

E_{pg} = gravitational potential energy (J)

m = mass of object (kg)

h = height above reference level (m)

g = acceleration due to gravity (m/s^2)

38. Conservation of Energy

In any isolated system, energy can be transformed from one kind to another, but the total amount of energy is constant (conserved):

$$E = E_k + E_p + E_e + ... = \text{constant}$$

Conservation of mechanical energy is given by

$$E_k + E_p = \text{constant}$$

39. Relativistic Energy

It is a consequence of relativity that the energy of a particle of rest mass m moving with speed v is given by

$$E = \frac{mc^2}{\sqrt{1 - \dfrac{v^2}{c^2}}}$$

where

m = rest mass of the body

MECHANICS
Dynamics

v = velocity of the body (m/s)

c = speed of light, $c = 3 \times 10^8$ m/s

$\dfrac{1}{\sqrt{1 - \dfrac{v^2}{c^2}}}$ = Lorentz factor

40. Power

Power is the rate at which work is done, or the rate at which energy is transformed from one form to another. Mathematically, it is computed using the following equation

$$P = \frac{W}{t}$$

where

P = power (W)

W = work (J)

t = time (s)

The standard metric unit of power is the watt (W). As is implied by the equation for power, a unit of power is equivalent to a unit of work divided by a unit of time. Thus, a watt is equivalent to Joule/second (J/s). Since the expression for work is

$$W = F \cdot s \, ,$$

MECHANICS
Statics

the expression for power can be rewritten as

$$P = F \cdot v$$

where

s = displacement (m)
v = speed (m/s)

41. Resolution of a Force

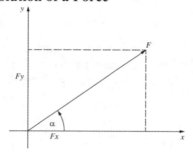

$$F_x = F \cos \alpha; \quad F_y = F \sin \alpha$$

$$F = \sqrt{F_x^2 + F_y^2}; \quad \tan \alpha = \frac{F_y}{F_x}$$

42. Moment of a Force about a Point 0

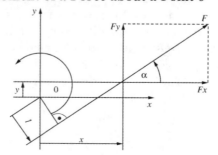

MECHANICS
Statics

$$M_0 = \pm Fl = F_y x - F_x y$$

43. Mechanical Advantage of Simple Machines

The mechanical advantage is the ratio of the force of resistance to the force of effort:

$$MA = \frac{F_R}{F_E}$$

where

MA = mechanical advantage

F_R = force of resistance (N)

F_E = force of effort (N)

44. The Lever

A lever consists of a rigid bar that is free to turn on a pivot, which is called a fulcrum.

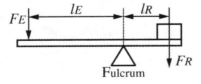

The law of simple machines as applied to levers is

$$F_R \cdot l_R = F_E \cdot l_E$$

MECHANICS
Statics

45. Wheel and Axle

A wheel and axle consist of a large wheel attached to an axle so that both turn together:

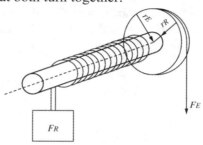

$$F_R \cdot r_R = F_E \cdot r_E$$

where

F_R = force of resistance (N)

F_E = force of effort (N)

r_R = radius of resistance wheel (m)

r_E = radius of effort wheel (m)

The mechanical advantage is

$$MA_{\text{wheel and axle}} = \frac{r_E}{r_R}$$

46. The Pulley

If a pulley is fastened to a fixed object, it is called a fixed pulley. If the pulley is fastened to the resistance to

MECHANICS
Statics

be moved, it is called a movable pulley. When one continuous cord is used, the ratio scales according to the number of strands holding the resistance in the pulley system.

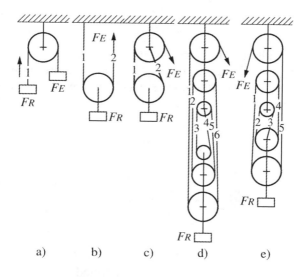

a)   b)   c)   d)   e)

The effort force equals the tension in each supporting strand. The mechanical advantage of the pulley is given by the formula

$$MA_{\text{pulley}} = \frac{F_R}{F_E} = \frac{nT}{T} = n$$

MECHANICS
Statics

where

T = tension in each supporting strand
n = number of strands holding the resistance

F_R = force of resistance (N)

F_E = force of effort (N)

47. The Inclined Plane

An inclined plane is a surface set at an angle from the horizontal and used to raise objects that are too heavy to lift vertically:

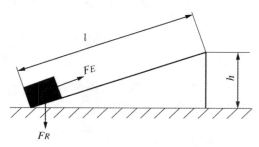

The mechanical advantage of an inclined plane is

$$MA_{\text{inclined plane}} = \frac{F_R}{F_E} = \frac{l}{h}$$

where

F_R = force of resistance (N)

F_E = force of effort (N)

MECHANICS
Statics

l = length of plane (m)
h = height of plane (m)

48. The Wedge

A wedge is a modification of the inclined plane. The mechanical advantage of a wedge can be found by dividing the length of either slope by the thickness of the longer end.

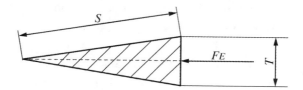

As with the inclined plane, the mechanical advantage gained by using a wedge depends on a corresponding increase in distance. The mechanical advantage is

$$MA = \frac{s}{T}$$

where

MA = mechanical advantage
s = length of either slope (m)
T = thickness of the longer end (m)

MECHANICS
Statics

---

49. The Screw

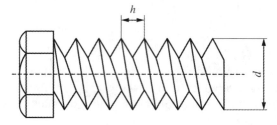

A screw is an inclined plane wrapped around a circle.
From the law of machines,

$$F_R \cdot h = F_E \cdot U_E$$

However, for advancing a screw with a screwdriver, the
mechanical advantage is:

$$MA_{screw} = \frac{F_R}{F_E} = \frac{U_E}{h}$$

where

F_R = force of resistance (N)

F_E = effort force (N)

h = pitch of screw

U_E = circumference of the handle of the screw

MECHANICS OF FLUIDS

The branch of mechanics called "mechanics of fluids" is concerned with fluids, which may be either liquids or gases. This topic studies various properties of fluids, such as velocity, pressure, density and temperature, as functions of space and time. Typically, liquids are considered to be incompressible, whereas gases are considered to be compressible.

This section of the book contains the most frequently used formulas and definitions related to the following:

- Hydrostatics
- Hydrodynamics

MECHANICS OF FLUIDS
Hydrostatics

---

1. Density

Density is the ratio of mass to volume:

$$\rho = \frac{m}{V}$$

where

ρ = density (kg/m^3)

m = mass (kg)

V = volume (m^3)

2. Viscosity

Viscosity is the measure of the internal friction between the molecules of liquid that resist motion across each other.

a) Dynamic viscosity:

Dynamic viscosity is a material constant which is a function of pressure and temperature:

$$\eta = f(p,t)$$

b) Kinematic viscosity:

$$v = \frac{\eta}{\rho}$$

where

v = kinematic viscosity $\left(m^2 / s\right)$
ρ = density (kg/m^3)
η = dynamic viscosity (Pa s)

$$1 \text{ Pa s} = \frac{kg}{m\,s} = \frac{N\,s}{m^2} = 10 \text{ P}$$

Viscosity measurements are expressed in "Pascal-seconds" (Pa s) or "milli-Pascal-seconds" (mPa s); these are units of the International System and are sometimes used in preference to the metric designations. But the most frequently used unit of viscosity measurement is the "poise" (P). (A material requiring a shear stress of one dyne per square centimeter to produce a shear rate of one reciprocal second has a viscosity of one poise, or 100 centipoise.)

One Pascal-second is equal to ten poise.
One milli-Pascal-second is equal to one centipoise.

3. Pressure of Solids

Pressure is force applied to a unit area:

$$p = \frac{F}{A}$$

where

p = pressure $(N/m^2$ or $lb/in^2)$

F = force applied (N or lb)

A = area, $(m^2 \text{ or } in^2)$.

$1 \ N/m^2 = 1 \ Pa$

4. Pressure in Liquids

Pressure in a liquid depends only on the depth and density of the liquid and not on the surface area. The pressure at any depth is, however, due not only to the weight of liquid above but to the pressure of air above the surface as well. The total pressure at a depth h is therefore given by the sum of these two pressures.

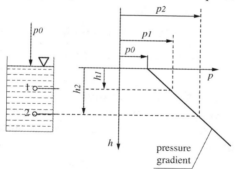

a) Pressure at a depth h_0:

The pressure at the free surface of the liquid ($h = 0$) is only the air pressure:

$$p_0 = p_a$$

b) Pressure at a depth h_1:

$$p_1 = p_0 + g\rho h_1$$

MECHANICS OF FLUIDS
Hydrostatics

---

c) Pressure at a depth h_2 :

$$p_2 = p_1 + g\rho(h_2 - h_1) = p_0 + g\rho h_2$$

where

p_1, p_2 = pressures at postions 1 and 2 (Pa)

h_1, h_2 = depths at positions 1 and 2 (m)

p_a = air pressure (Pa)

p_0 = pressure on free surface of
the liquid (Pa)

ρ = density of the liquid (kg/m^3)

g = acceleration due to gravity (m/s^2)

5. Force Exerted by Liquids

a) Force on a horizontal surface:
The force exerted by a liquid on a horizontal surface is
given by the formula

$$F = g\rho h A_h$$

where

A_h = area of horizontal surface (m^2)

h = depth of the surface in the liquid (m)

ρ = density of the liquid (kg/m^3)

g = acceleration due to gravity (m/s^2)

MECHANICS OF FLUIDS
Hydrostatics

---

b) Force on a vertical surface:
The force on a vertical surface is found by using half the vertical height and is given by the formula

$$F_s = \frac{1}{2} g \rho h_a A_s$$

where

A_s = area of the side or vertical surface (m^2)

h_a = average depth of the surface in the liquid (m)

ρ = density of the liquid (kg/m^3)

g = acceleration due to gravity (m/s^2)

6. Pascal's Principle
Pressure exerted on an enclosed liquid is transmitted equally to every part of the liquid and to the walls of the container. Pascal's principle is important in understanding hydraulics, which is the study of the transfer of forces through fluids.

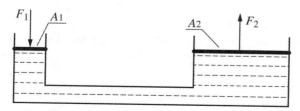

MECHANICS OF FLUIDS
Hydrostatics

$$p = \frac{F_1}{A_1} = \frac{F_2}{A_2}$$

where

A_1, A_2 = areas of small and large surfaces (m^2)

F_1, F_2 = applied and upward forces (N)

7. Archimedes' Principle

Any object placed in a fluid apparently loses weight equal to the weight of the displaced fluid.

For water, which has a density of $\rho_w = 1$ g/cm^3, this provides a convenient way to determine the volume of an irregularly shaped object and thus determine its density:

$$m_o - m_{app} = \rho_w \cdot V_o$$

where

m_o = mass of object (kg)

m_{app} = apparent mass when submerged (kg)

V_o = volume of object (m^3)

ρ_w = density of water (kg/m^3)

MECHANICS OF FLUIDS
Hydrostatics

8. Buoyant Force

When a rigid object is submerged in a fluid, there exists a buoyant force (an upward force) on the object that is equal to the weight of the fluid displaced by the object. This force is given by the equation

$$F_b = \rho g V$$

where

F_b = buoyant force (N)

ρ = density of the liquid (kg/m^3)

g = acceleration due to gravity (m/s^2)

V = volume of submerged object (m^3)

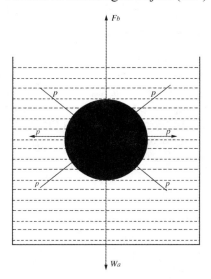

MECHANICS OF FLUIDS
Hydrostatics

The net force on the object is given by

$$F_n = F_b - W_o = \rho_f \cdot V_s \cdot g - \rho_o \cdot V_o \cdot g$$

where

F_n = net force on object (N)

F_b = buoyant force (N)

W_o = weight of the object (kg)

ρ_f = density of the fluid (kg/m^3)

V_s = volume submerged (m^3)

ρ_o = density of the object (kg/m^3)

V_o = volume of the object (m^3)

g = acceleration due to gravity (m/s^2)

When an object is floating, the net force on it will be zero. This happens when the volume of the object that is submerged displaces an amount of liquid whose weight is equal to the weight of the object. A ship made of steel can float because it can displace more water than it weighs.

MECHANICS OF FLUIDS
Hydrodynamics

9. Flow Rate

The flow rate of a fluid is the volume of fluid flowing past a given point in a pipe per unit time:

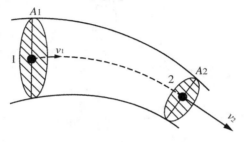

$$Q = A_1 \cdot v_1 = A_2 \cdot v_2 = \text{constant}$$

where

Q = flow rate (m^3 / s)

v_1, v_2 = flow velocities at point 1 and point 2 (m/s)

A_1, A_2 = cross-sectional areas at sections 1 and 2 (m²)

10. Conservation of Mass

The rate at which fluid mass goes into a system is equal to the rate of accumulation plus the rate at which mass goes out. At steady (lamellar) state, the rate of accumulation is zero; therefore

$$A_1 v_1 \rho_1 = A_2 v_2 \rho_2 = Av\rho$$

MECHANICS OF FLUIDS
Hydrodynamics

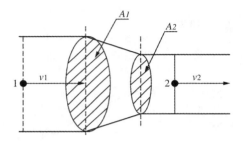

where

A_1, A_2 = areas of the pipe cross-section at point 1 and point 2 (m^2)

v_1 = fluid velocity at point 1 (m/s)

v_2 = fluid velocity at point 2 (m/s)

ρ_1 = density of fluid at point 1 (kg/m^3)

ρ_2 = density of fluid at point 2 (kg/m^3)

11. Bernoulli's Equation

The Bernoulli equation is a statement derived from conservation of energy and work-energy ideas that come from Newton's Laws of Motion.

Bernoulli's equation is based on the concept that points 1 and 2 lie on a streamline, the fluid has constant density, the flow is steady, and there is no friction.

MECHANICS OF FLUIDS
Hydrodynamics

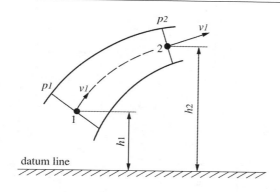

$$p_1 + h_1\rho g + \frac{1}{2}\rho v_1^2 = p_2 + h_2\rho g + \frac{1}{2}\rho v_2^2$$

where

$p_1 =$ fluid pressure at point 1 (Pa)

$p_2 =$ fluid pressure at point 2 (Pa)

$v_1 =$ fluid velocity at point 1 (m/s)

$v_2 =$ fluid velocity at point 2 (m/s)

$h_1 =$ elevation at point 1 (m)

$h_2 =$ elevation at point 2 (m)

$g =$ acceleration due to gravity (m/s^2)

TEMPERATURE AND HEAT

Thermodynamics is a branch of physics. It is the study of the effects of work, heat, and energy on systems. Heat is a form of energy transferred from one body or system to another as a result of a difference in temperature. The energy associated with the motion of atoms or molecules is capable of being transmitted through solid and fluid media by conduction, through fluid media by convection, and through empty space by radiation.

Temperature is the specific degree of hotness or coldness of a body or an environment. It is usually measured with a thermometer or other instrument having a scale calibrated in units (degrees).

This section contains the most frequently used formulas, rules, and definitions relating to:

- Thermal Variables of State
- Heating of Solid and Liquid Bodies
- Changes of State
- Gas Laws
- Laws of Thermodynamics

TEMPERATURE AND HEAT
Thermal Variables of State

1. Pressure

The pressure of a system is defined as the force exerted by the system per unit area of its boundaries. This is the definition of absolute pressure. A state of pressure means $p_g > 0$, and a vacuum means $p_g < 0$. Thus, absolute pressure can be expressed as

$$p = p_g + p_0$$

where

p = absolute pressure (Pa)

p_g = gauge pressure (Pa)

p_0 = atmospheric pressure (Pa)

2. Temperature

Basically, temperature is a measure of the hotness or coldness of an object. There are four basic temperature scales: Celsius (°C), Kelvin (K), Fahrenheit (°F), and Rankine (°R).

Boiling point of water	100°C	373 K	212°F	672°R
Freezing point of water	0°C	273 K	32°F	492°R
Absolute zero	−273°C	0 K	−460°F	0°R

TEMPERATURE AND HEAT
Thermal Variables of State

The Kelvin scale is closely related to the Celsius scale:
$$t_K = t_C + 273°$$

The Rankine scale is closely related to the Fahrenheit scale:
$$t_R = t_F + 460°$$

The relationship between Celsius temperatures and Fahrenheit temperatures is given by
$$t_C = \frac{5}{9}\left(t_F - 32°\right); \quad t_F = \frac{9}{5}t_C + 32°$$

3. Density
Density is a measurement of mass per unit of volume:
$$\rho = \frac{m}{V}$$

where

ρ = object's density (kg/m^3)

m = object's total mass (kg)

V = object's total volume (m^3)

4. Specific Volume
Specific volume is the volume per unit mass or the inverse of density:
$$v = \frac{V}{m} = \frac{1}{\rho}$$

---

where

v = specific volume (m^3/kg)

ρ = object's density (kg/m^3)

m = object's total mass (kg)

V = object's total volume (m^3)

5. Molar Mass

Molar mass is the mass of one mole of a substance.

a) Mass of one molecule:

$$m_M = M_r \cdot u$$

where

u = unified atomic mass ($u = 1.66 \times 10^{-27}$ kg)

M_r = relative molecular mass

The relative molecular mass of a substance is equal to the sum of the relative atomic masses of its constituent atoms.

b) Molar mass of a substance:

$$M = \frac{m}{n} = M_r \cdot N_A$$

where

m = total mass of the substance (g)

n = number of moles of the substance (mol)

N_A = Avogadro's number (mol^{-1})

TEMPERATURE AND HEAT
Heating of Solid and Liquid Bodies

---

6. Molar Volume

The molar volume is the volume occupied by one mole of ideal gas at standard temperature and pressure (STP).

a) Standard temperature:
$$T_0 = 273.15K = 0^{\circ}C$$

b) Standard pressure:
$$p_0 = 101325 \, Pa = 1.03 \, bar$$

c) Molar volume at STP:
$$V_m = 2.24 \times 10^{-2} \, m^3 mol^{-1}$$

d) Volume of a gas
$$V = nV_m$$

7. Heat

Heat is the energy that flows spontaneously from a higher temperature object to a lower temperature object through random interactions between their atoms. Heat is expressed as

$$Q = mc(T_2 - T_1)$$

where

Q = heat thermal energy (J)

m = cooler object's mass (kg)

c = specific heat (J/kg K)

T_2 = temperature of the hotter object (K)

T_1 = temperature of the cooler object (K)

TEMPERATURE AND HEAT
Heating of Solid and Liquid Bodies

---

8. Specific Heat
The specific heat is the amount of heat per unit mass required to raise the temperature by one degree Celsius:

$$c = \frac{Q}{m\Delta T}$$

9. Heat Conduction
The total amount of heat transfer between two plane surfaces is given by the equation

$$Q = \frac{kAt(T_2 - T_1)}{l}$$

where

Q = heat transferred (J or Btu)

k = thermal conductivity $\left(\text{J/s m }^{\circ}\text{C}\right)$

A = plane area (m^2)

l = thickness of barrier (m)

T_2 = temperature of the hotter side (K)

T_1 = temperature of the cooler side (K)

10. Expansion of Solid Bodies
a) Linear expansion:
The amount by which a solid expands can be found in the formula

$$\Delta l = \alpha\, l\Delta T$$

where

Δl = change in length (m)

l = original length (m)

α = coefficient of linear expansion (m / °C)

ΔT = change in temperature

b) Area and volume expansion:

To allow for this expansion, the following formulas are used:

$$\Delta A = 2\alpha A \Delta T$$

$$\Delta V = 3\alpha V \Delta T$$

where

A = original area $\left(m^2\right)$

V = original volume $\left(m^3\right)$

11. Expansion of Liquids

The formula for volume expansion of liquids is

$$\Delta V = \beta V \Delta T$$

where V = original volume $\left(m^3\right)$

β = coefficient of volume expansion for the liquid

TEMPERATURE AND HEAT
Changes of State

---

12. Expansion of Water
The most common liquid, water, does not behave like other liquids. Above about 4°C, water expands as the temperature rises, as we would expect. Between 0 and about 4°C, however, water *contracts* with increasing temperature. Thus, at exactly 3.98°C, the density of water passes through a maximum. At all other temperatures, the density of water is less than this maximum value.

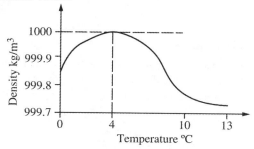

13. Fusion
The change of state from solid to liquid is called fusion or melting. The change from the liquid to the solid state is called freezing or solidification. The heat of fusion L_f is the quantity of heat energy required to convert one mass unit of solid to liquid form:

$$L_f = \frac{Q}{m}$$

where

Q = quantity of heat (J)
m = mass needed (kg)

---

14. Vaporization
The change of state from a liquid to a gaseous or vaporous state is called vaporization.

The heat of vaporization L_v is the heat required to vaporize one mass unit of a substance at its normal boiling point:

$$L_v = \frac{Q}{m}$$

15. Equation of State
The equation of state of a gas in thermal equilibrium relates the pressure, the volume, and the temperature of a gas. All gases have the same equation of state, called the ideal gas law:

$$pV = NkT = nRT$$

where

N = number of molecules in the gas
n = number of moles of the gas (mol)
T = Kelvin temperature of the gas (K)
p = pressure (Pa)
V = volume $\left(m^3\right)$
k = Boltzmann's constant $(k = 1.38 \times 10^{-23} \text{ J/K})$
R = universal gas constant $(R = 8.314 \text{ J/mol} \cdot \text{K})$

TEMPERATURE AND HEAT
Gas Laws

The ratio

$$N_A = \frac{R}{k} = 6.022 \times 10^{23} \text{ mol}^{-1}$$

is Avogadro's number, which is the number of molecules in a mole.

16. The Charles Law for Temperature

If the pressure on a gas is constant, p = constant, the volume is directly proportional to its absolute temperature:

$$\frac{V_1}{T_1} = \frac{V_2}{T_2} \quad \text{or} \quad V_1 T_2 = V_2 T_1$$

where

V_1 = original volume (m^3)

T_1 = original temperature (K)

V_2 = final volume (m^3)

T_2 = final temperature (K)

17. Boyle's Law for Pressure

If the temperature of the gas is constant, T = constant, the volume and pressure are inversely proportional:

$$p_1 V_1 = p_2 V_2$$

TEMPERATURE AND HEAT
Gas Laws

---

where

V_1 = original volume (m^3)

p_1 = original pressure (Pa)

V_2 = final volume (m^3)

p_2 = final pressure (Pa)

18. Gay-Lussac's Law for Temperature

The pressure of a given mass of gas is directly proportional to the Kelvin temperature if the volume is kept constant:

$$\frac{p_1}{T_1} = \frac{p_2}{T_2}$$

where

p_1 = original pressure (Pa)

p_2 = final pressure (Pa)

T_1 = original temperature (K)

T_2 = final temperature (K)

19. Dalton's Law of Partial Pressures

At constant volume and temperature, the total pressure (p_T) exerted by a mixture of gases is equal to the sum of the partial pressures:

$$p_T = p_1 + p_2 + p_3 + ... + p_n$$

where p_T = total pressure (Pa)

$p_1 + p_2 + p_3 + ... + p_n$ = partial pressures (Pa)

20. Combined Gas Law

Most of the time, it is very difficult to keep pressure or temperature constant. To keep these parameters constant, the best solution is to combine Charles' law and Boyle's law as follows:

$$\frac{p_1 V_1}{T_1} = \frac{p_2 V_2}{T_2}$$

21. The First Law of Thermodynamics

The first law of thermodynamics is often called the law of conservation of energy. This law states that energy can be transformed from one kind of matter to another in many forms. However, it cannot be created nor destroyed.

The first law of thermodynamics defines internal energy (E) as equal to the heat transfer (Q) *into* a system and the work (W) done by the system.

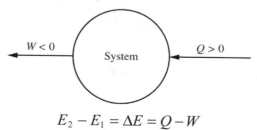

$$E_2 - E_1 = \Delta E = Q - W$$

TEMPERATURE AND HEAT
Laws of Thermodynamics

where

ΔE = change in internal energy
Q = heat added into the system
W = work done by the system

22. The Second Law of Thermodynamics

In physics, the second law of thermodynamics, in its many forms, is a statement about the quality and direction of energy flow, and it is closely related to the concept of entropy. This law says that heat can never pass spontaneously from a colder to a hotter body. As a result of this fact, natural processes that involve energy transfer must have one direction, and all natural processes are irreversible.

a) Entropy:

Thermodynamic entropy (S) is a measure of the amount of energy in a physical system that cannot be used to do work. It is a state variable whose change is defined for a reversible process via the temperature T and the heat absorbed Q. The entropy change is

$$\Delta S = \frac{Q}{T}$$

where

ΔS = entropy change (J/K)
Q = heat absorbed (J)
T = temperature (K)

TEMPERATURE AND HEAT
Laws of Thermodynamics

The importance of the entropy function is exhibited in the second law of thermodynamics.

In any process, the total entropy of the system and its surroundings either increases or (in a reversible process) does not change.

b) Heat engines and refrigerators:

A heat engine is a device or system that converts heat into work. The efficiency of a cyclic heat engine is

$$\eta = \frac{W}{Q_h} = 1 - \frac{Q_c}{Q_h}$$

where

Q_h = heat absorbed per cycle from the higher temperature (J)

Q_c = heat rejected per cycle to the lower temperature (J)

W = work carried out per cycle (J)

The most efficient heat engine cycle is the Carnot cycle, consisting of two isothermal processes and two adiabatic processes.

TEMPERATURE AND HEAT
Laws of Thermodynamics

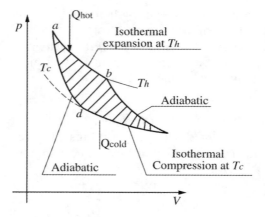

This maximum thermal efficiency is

$$\eta = 1 - \frac{T_c}{T_h} \quad (T_h > T_c), \text{ but also}$$

$$\eta = 1 - \frac{Q_{cold}}{Q_{hot}}$$

23. The Third Law of Thermodynamics

The third law of thermodynamics states that the entropy of a system at zero absolute temperature is a well-defined constant.

Absolute zero = 0 K = −273.15 $^{\circ}C$

ELECTRICITY AND MAGNETISM

Electricity is electrical charge. Franklin, Faraday, Thompson, Einstein, Tesla, and many other historical scientists used the word "electricity" in this way, stating that an electric current is a flow of electricity.

Magnetism is a force that acts at a distance and is caused by a magnetic field. This force strongly attracts ferromagnetic materials such as iron, nickel and cobalt.

With magnets, a magnetic force strongly attracts the opposite pole of another magnet and repels the like pole. A magnetic field is both similar to and different from an electric field.

This section contains the most frequently used formulas, rules, and definitions regarding the following:

- Electrostatics
- Direct Current
- Magnetism
- Alternating Current

371

ELECTRICITY AND MAGNETISM
Electrostatics

---

1. Coulomb's Law

The force between two point charges Q_1 and Q_2 is directly proportional to the product of their magnitudes and inversely proportional to the square of the distance separating them, r.

In equation form, Coulomb's law is

$$F = \frac{kQ_1Q_2}{r^2}$$

where

F = force of attraction or repulsion (N)

k = constant $\left(k = 8.99 \times 10^9 \, \text{N m}^2/\text{C}^2 \text{ for air}\right)$

Q_1, Q_2 = size of charges in coulombs (C)

r = distance between the charges (m)

2. Electric Fields

Electric field strength is a vector quantity having both magnitude and direction.

The magnitude of the electric field of a point charge is simply defined as the force per charge exerted on a test charge:

$$E = \frac{F}{q}$$

where

E = electric field strength (N/C)

ELECTRICITY AND MAGNETISM
Electrostatics

---

q = quantity of charge of the test charge (C)
F = electric force (N).

When applied to two charges, the source charge Q and the test charge q, the formula for field strength force can be written as

$$E = \frac{kQ}{r^2}$$

a) The principle of superposition for electric fields: The total electric field at any point, generated by a distribution of charges $q_1, q_2, q_3, ..., q_n$, is found by adding the fields independently established at that point by the individual charges:

$$E_{total} = E_1 + E_2 + E_3 + ... + E_n$$

3. Electric Flux

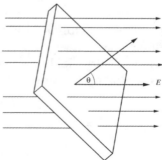

The electric flux is the product of the components of the electric field that are perpendicular to a given surface, times the surface area:

ELECTRICITY AND MAGNETISM
Electrostatics

$$\Phi_E = EA\cos\theta$$

where

Φ_E = electric flux $\left(\text{N m}^2 / \text{C}\right)$

θ = angle between field and area vector

A = area vector $\left(\text{m}^2\right)$

E = electric field (N/C)

4. Gauss's Law

The electric flux through any closed surface is equal to the charge enclosed by that surface divided by a constant ε_0:

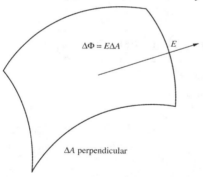

$$\Phi_E = \oint \vec{E}d\vec{A} = \frac{q_{inc}}{\varepsilon_0}$$

where

Φ_E = total electric flux

ε_0 = permittivity of free space constant

$= 8.854 \times 10^{-12} \left(\text{C}^2 / \text{Nm}^2\right)$

q_{inc} = sum of all the enclosed charges

ELECTRICITY AND MAGNETISM
Electrostatics

These equations apply in a vacuum and for the most part also in air.

5. Electric Potential

Electric potential can be expressed as potential energy per unit charge.

The electric potential V at a distance r from a charge q is

$$V = k\frac{q}{r}$$

where

V = electrical potential (V)
k = constant $\left(k = 8.99 \times 10^9 \, \text{N m}^2/\text{C}^2\right)$
q = charge (C)
r = distance (m)

a) Principle of superposition of electric potential: When more than one charge is present, the electric potential at a given point is the algebraic sum of the potentials due to each of the charges present. The electric potential V at any point is given by

$$V = V_1 + V_2 + V_3 + \dots + V_n = k\sum \frac{q_i}{r_i}$$

where

q_i = charge
r_i = distance of the charge
V_n = potentials due to n different charges

ELECTRICITY AND MAGNETISM
Electrostatics

---

6. Electric Potential Energy

The electric potential can also be defined as the electric potential energy per unit charge. Hence,

$$V = \frac{U}{q} = \frac{W}{q}$$

where

U = magnitude of electric potential energy
W = work done
q = charge

7. Capacitance

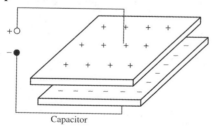

Capacitor

Capacitance is a measure of the amount of stored electric charge for a given electric potential:

$$C = \frac{Q}{V}$$

where

C = capacitance (F)
Q = total electric charge (C)
V = electric potential (V)

ELECTRICITY AND MAGNETISM
Electrostatics

---

8. Capacitor

The capacitance of a capacitor can be calculated by the following formula:

$$C = \varepsilon_0 \varepsilon_r \frac{A}{d}$$

where

C = capacitance (F)

ε_0 = permittivity of free space (F/m)

ε_r = dielectric constant of the insulator (F/m)

A = area of each electrode plate $\left(m^2\right)$

d = distance between the electrodes $\left(m^2\right)$

a) Capacitances in parallel

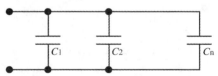

The equivalent capacitance of capacitors connected in parallel is

$$C_{eq} = C_1 + C_2 + ... + C_n$$

b) Capacitances in series

ELECTRICITY AND MAGNETISM
Direct Current

The equivalent capacitance of capacitors connected in series is

$$\frac{1}{C_{eq}} = \frac{1}{C_1} + \frac{1}{C_2} + ... + \frac{1}{C_n}$$

c) Energy

The energy stored in a capacitor is equal to the *work* done to charge it up

$$W_{sto} = \frac{1}{2}CV^2$$

where

W_{sto} = energy stored in capacitor (J)

C = capacitance (F)

V = electric potential (V)

9. Electric Current

The rate of flow of electrons through a conductor from a negatively charged area to one that has a positive charge is called direct current. Thus,

$$I = \frac{Q}{t}$$

where

I = current (A)

Q = charge (C)

t = time (s)

ELECTRICITY AND MAGNETISM
Direct Current

---

10. Current Density
The magnitude of a current's density is the current through a unit area perpendicular to the flow direction. Thus,

$$J = \frac{I}{A}$$

where

J = current density (A/m^2)

A = conductor's cross-section area (m^2)

I = current (A)

11. Potential Difference
The electric potential difference (V) is the work done per unit charge as a charge is moved between two points a and b in an electric field:

$$V_a - V_b = V = \frac{W_{ab}}{Q}$$

where

V = electric potential difference (V)

W_{ab} = work done as a charge moved between points a and b (J)

Q = charge (C)

12. Resistance
Resistance is the feature of a material that determines the flow of electric charge:

$$R = \rho \frac{l}{A}$$

where

R = resistance (Ω)
l = length (m)
A = cross-section area (m^2)
ρ = resistivity, a constant, which depends on the
     type of material ($\Omega \cdot$m)

Very often one specifies, instead of ρ, the conductivity

$$\sigma = \frac{1}{\rho}$$

where

σ = conductivity (S/m)

13. Ohm's Law

The current I in a "resistor" is very nearly proportional
to the difference V in electric potential between the ends
of the resistor. This proportionality is expressed by
Ohm's law:

$$V = IR \quad \text{or} \quad I = \frac{V}{R}$$

where

I = current through the resistor (A)
V = potential difference (V)
R = resistance (Ω)

382

ELECTRICITY AND MAGNETISM
Direct Current

---

14. Series Circuits

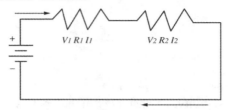

a) Potential difference
The total potential difference is the sum of the potential differences of the individual components:

$$V = V_1 + V_2 + \cdots + V_n$$

b) Resistance
The total resistance is equal to the sum of the resistances of the components:

$$R = R_1 + R_2 + \cdots + R_n$$

c) Current
The total current is equal in every component.

$$I = I_1 = I_2 = \cdots = I_n$$

15. Parallel Circuits

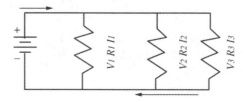

a) Potential difference
The total potential difference is equal in every component.

$$V = V_1 = V_3 = V_3 = \cdots = V_n$$

b) Resistance
The total resistance of a parallel circuit is not equal to the sum of the resistances. The total resistance in a parallel circuit:

$$\frac{1}{R} = \frac{1}{R_1} + \frac{1}{R_2} + \frac{1}{R_3} + \cdots + \frac{1}{R_n}$$

c) Current
The total current is equal to the sum of the currents in the individual components:

$$I = I_1 + I_2 + I_3 + \cdots + I_n$$

16. Series-Parallel Circuit

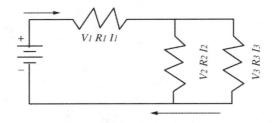

Many circuits have both series and parallel components.

ELECTRICITY AND MAGNETISM
Direct Current

---

a) Potential difference

The total potential difference is the potential difference of the series circuit plus the potential difference of parallel circuits.

$$V = V_1 + V_2 = V_1 + V_3$$

b) Resistance

The total resistance is the resistance of the series circuit plus the resistance of the parallel circuits.

$$R = R_1 + \frac{R_2 R_3}{R_2 + R_3}$$

c) Current

The total current is equal to the current of the series circuit and to the sum of the currents of the parallel circuits.

$$I = I_1 = I_2 + I_3$$

17. Joule's Law

a) Work

The "work" or heat energy produced by a resistor is

$$W = I^2 Rt = \frac{V^2}{R} t$$

where

W = work or heat energy (J)

ELECTRICITY AND MAGNETISM
Direct Current

I = current (A)
R = resistance (Ω)
V = potential difference (V)
t = time (s)

b) Power

Electrical power is defined as the rate of doing work. The power consumption of a resistor is

$$P = VI = I^2 R = \frac{V^2}{R}$$

where

P = power (W)
I = current (A)
R = resistance (Ω)
V = potential difference (V).

18. Kirchhoff's Junction Law

For a given junction or node in a circuit, the sum of the currents entering equals the sum of the currents leaving it. In other words, the algebraic sum of all the currents in the junction is zero (as, for example, $I_1 + I_2 = I_3$). In this case, a current going out of the junction is counted as negative.

ELECTRICITY AND MAGNETISM
Direct Current

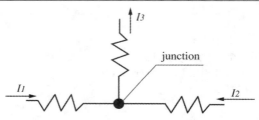

That is, at any junction,

$$\sum_{j=1}^{n} I_j = 0$$

19. Kirchhoff's Loop Law

The algebraic sum of the potential changes around any complete loop in the network is zero, so the sum of the voltage drops equals the voltage source.

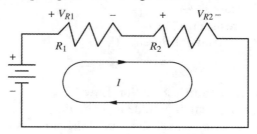

In this example,

$$V = V_{R1} + V_{R2}$$

That is, over any complete loop,

$$\sum_{loop} V = 0$$

20. Resistors

Electrical components called *resistors* are used to limit or set the current in a circuit with a given voltage, or to the set voltage for a given current. (A circuit *element* is an idealization of an actual electronic part, or *component*.) Resistors are usually marked with at least three color bands that indicate their resistance in units of ohms Ω. For 5% tolerance resistors, the first two bands are the first two significant digits of the value, and the third band is the number of zeros to be added to the first two digits. A final band of gold (5%) or silver (10%) indicates the tolerance. For 1% resistors, the first three bands are the first three digits; the fourth is the multiplier. The color code is:

BLACK 0, BROWN 1, RED 2, ORANGE 3, YELLOW 4, GREEN 5, BLUE 6, VIOLET 7, GRAY 8, WHITE 9.

21. Internal Resistance

A cell has resistance within itself, which opposes the movement of electrons. This is called the internal resistance. The voltage applied to the external circuit is, then,

$$V = E - I \cdot r$$

where

V = voltage applied to circuit (V)
E = potential difference across a source (V)

I = current through cell (A)

r = internal resistance of cell (Ω)

22. Magnetic Forces on Moving Charges

A magnetic field is an entity produced by moving electric charges exerting a force on other moving charges. The following equation describes this force:

$$F = qvB \sin\theta$$

where

F = force (N)

q = electric charge (C)

v = velocity of the charge (m/s)

B = strength of the magnetic field (T)

θ = smaller angle between the vectors v and B

$$1\,\text{T} = 1\frac{\text{N}\cdot\text{s}}{\text{C}\cdot\text{m}} = 1\frac{\text{N}}{\text{A}\cdot\text{m}} = \frac{\text{V}\cdot\text{s}}{\text{m}^2}$$

23. Force on a Current-Carrying Wire

If, instead of a moving charge such as an electron or proton, there is electric current going through a wire, the total force is the result of the current and the magnetic field:

$$F = B \cdot I \cdot L \sin \theta$$

where

L = length of the wire through the
magnetic field (m)

24. Magnetic Field of a Moving Charge

The magnetic field near a long current-carrying wire
is circular about the wire and is given by

$$B = \frac{\mu_0 I}{2\pi r}$$

where

B = strength of the magnetic field (T)

I = current through the wire (A)

r = perpendicular distance from the center of the
wire (m)

μ_0 = permeability of empty space

$= 4\pi \times 10^{-7}$ (H/m)

The henry (H) is the unit of inductance.

$$1 \text{ H} = 1\frac{\text{N} \cdot \text{s}^2 \cdot \text{m}}{\text{C}^2} = 1\frac{\text{Wb}}{\text{A}}$$

25. Magnetic Field of a Loop

For a long current-carrying coil that is tightly
turned (a solenoid), the magnetic field strength at
its center is

ELECTRICITY AND MAGNETISM
Alternating Current

$$B = \mu_0 In$$

where

> n = number of turns per unit length of solenoid (turns/m)

> B = magnetic field in the region at the center of the solenoid (T)

μ_0 = permeability constant ($\mu_0 = 4\pi \times 10^{-7}$ H/m)

I = current through the solenoid (A)

26. Faraday's Law

If the magnetic flux changes by $d\Phi$ in a time dt, then the induced electromotive force is given by

$$\varepsilon = -N \frac{d\Phi}{dt}$$

where

> ε = induced electromotive force (V)

$d\Phi$ = differential of change of the magnetic flux (Wb)

dt = differential change in time (s)

N = numbers of turns per loop.

> The minus sign means that the magnetic field produced by the induced current opposes the external field produced by the magnet.

27. Properties of Alternating Current

An alternating current (AC) is an electrical current in which the magnitude and direction of the current vary

ELECTRICITY AND MAGNETISM
Alternating Current

cyclically, as opposed to direct current, in which the direction of the current stays constant. The usual wave form of an AC power circuit is a sine wave, as this results in the most efficient transmission of energy.

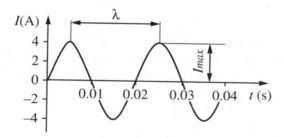

28. Period
The time required to complete one cycle of a waveform is called the period of the wave:

$$t = \frac{1}{f}$$

29. Frequency
The number of complete cycles of alternating current or voltage completed each second is referred to as the frequency:

$$f = \frac{1}{t}$$

30. Wavelength
The distance traveled by the sine wave during one period is referred to as the wavelength:

$$\lambda = \frac{c}{f}$$

where
c = speed of light, $c = 3.00 \times 10^8$ (m/s)

31. Instantaneous Current and Voltage

Instantaneous current is the current at any instant of time. Instantaneous voltage is the voltage at any instant of time:

$$i = I_{max} \sin\theta, \quad e = E_{max} \sin\theta$$

where
i = instantaneous current (A)
I_{max} = maximum instantaneous current (A)
e = instantaneous voltage (V)
E_{max} = maximum instantaneous voltage (V)
θ = angle measured from beginning of cycle

32. Effective Current and Voltage

A direct measurement of AC is difficult because it is constantly changing. The most useful value of AC is based on its heating effect and is called its effective value. The effective value of an AC is the number of amperes that produces the same amount of heat in a resistance as an equal number of amperes of a steady direct current. The equations for effective current and voltage are

$$I_{\text{eff}} = 0.707 I_{\max}$$
$$E_{\text{eff}} = 0.707 E_{\max}$$

where

$I_{\text{eff}}, E_{\text{eff}}$ = effective value of current and voltage

$I_{\max}, E_{\max}$ = maximum or peak current and voltage

33. Maximum Current and Voltage

When I_{eff} or E_{eff} is known, you can find $I_{\max}$ and $E_{\max}$ by using the formulas

$$I_{\max} = 1.41 I_{\text{eff}}$$
$$E_{\max} = 1.41 E_{\text{eff}}$$

34. Ohm's Law for AC Current Containing Only Resistance

Many AC circuits contain resistance only. The rules for these circuits are the same rules that apply to DC circuits. The Ohm's Law formula for an AC circuit is

$$I = \frac{E}{R}$$

NOTE: Do not mix AC values. When you solve for effective values, all the values you use in the formula must be effective values.

35. AC Power

When AC circuits contain only resistance, power is found in the same way as in DC circuits:

$$P = I^2 R = EI = \frac{E^2}{R}$$

36. Changing Voltage with Transformers

If we assume no power loss between primary and secondary coils, we have the following equation:

$$\frac{E_P}{E_S} = \frac{N_P}{N_S}$$

where

E_P = primary voltage (V)
E_S = secondary voltage (V)
N_P = number of turns in the primary coil
N_S = number of turns in the secondary coil

The relationship between primary and secondary currents is

$$\frac{I_S}{I_P} = \frac{N_P}{N_S}$$

where

I_S = current in secondary coil (A)
I_P = current in primary coil (A)
N_P = number of turns in primary coil
N_S = number of turns in secondary coil

37. Inductive Reactance

The opposition to AC current flow in an inductor is called inductive reactance and is measured in ohms:

$$X_L = 2\pi f L$$

where

X_L = inductive reactance (Ω)
f = frequency of the AC voltage (Hz)
L = inductance (H)

The current in a circuit that has only an AC voltage source and inductor is given by

$$I = \frac{E}{X_L}$$

where

I = current (A)
E = voltage (V)
X_L = inductive reactance (H)

ELECTRICITY AND MAGNETISM
Alternating Current

---

38. Inductance and Resistance in Series

The effect of both the resistance and the inductance on a circuit is called the impedance:

$$Z = \sqrt{R^2 + X_L^2} = \sqrt{R^2 + (2\pi f L)^2}$$

where

Z = impedance (Ω)

R = resistance (Ω)

X_L = inductive reactance (Ω)

f = frequency of the AC voltage (Hz)

L = inductance (H)

a) Phase angle

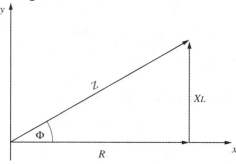

The phase angle is given by

$$\tan \phi = \frac{X_L}{R}$$

The resistance is always drawn as a vector pointing in the positive x-axis, and inductive reactance is drawn as a vector pointing in the direction of the positive y-axis.

ELECTRICITY AND MAGNETISM
Alternating Current

b) Ohm's law

In general, Ohm's law cannot be applied to alternating-current circuits since it does not consider the reactance which is always present in such circuits:

$$I = \frac{E}{Z}$$

where

I = current (A)
Z = impedance (Ω)
E = voltage (V)

39. Capacitance

The effect of a capacitor on a circuit is inversely proportional to frequency and is measured as capacitive reactance, which is given by

$$X_C = \frac{1}{2\pi f C}$$

where

X_C = capacitive reactance (Ω)
f = frequency (Hz)
C = capacitance (F)

40. Capacitance and Resistance in Series

The impedance of the circuit measures the combined effect of resistance and capacitance in series

ELECTRICITY AND MAGNETISM
Alternating Current

$$Z = \sqrt{R^2 + X_C^2} = \sqrt{R^2 + \left(2\pi f C\right)^2}$$

where

Z = impedance (Ω)
R = resistance (Ω)
X_C = inductive reactance (Ω)
f = frequency of the AC voltage (Hz)
C = capacitance (F)

a) Current
The formula for current is given by Ohm's law:

$$I = \frac{E}{Z}$$

where

I = current (A)
Z = impedance (Ω)
E = voltage (V)

b) Phase angle

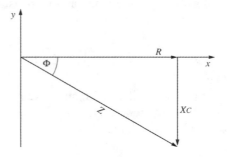

The phase angle gives the amount by which the voltage lags behind the current:

$$\tan \phi = \frac{X_C}{R}$$

---

41. Capacitance, Inductance, and Resistance in Series

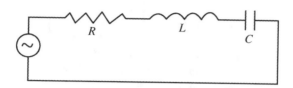

The impedance of a circuit containing resistance, capacitance, and inductance in series can be calculated by the equation

$$Z = \sqrt{R^2 + \left(X_L - X_C\right)^2}$$

where

Z = impedance (Ω)

R = resistance (Ω)

X_C = capacitive reactance (Ω)

X_L = inductive reactance (Ω)

a) Phase angle:

The phase angle is given by the following formula:

$$\tan \phi = \frac{X_L - X_C}{R}$$

b) Current

The current in this type of circuit is given by

ELECTRICITY AND MAGNETISM
Alternating Current

$$I = \frac{E}{Z} = \frac{E}{\sqrt{R^2 + \left(2\pi f L - \frac{1}{2\pi f C}\right)^2}}$$

c) Frequency

The resonant frequency occurs when $X_L = X_C$. This frequency can be calculated by

$$f = \frac{1}{2\pi\sqrt{LC}}$$

42. Power in AC Circuits

When the current and voltage are in phase, the power can be stated as

$$P = EI$$

where

P = power (W)
E = voltage (V)
I = current (A)

a) Apparent power

If current and voltage are not in phase, the resultant product of current and voltage is apparent power (S).

$$S = E \cdot I = \sqrt{P^2 + Q^2} = I^2 Z$$

b) Real power

ELECTRICITY AND MAGNETISM
Alternating Current

Real power or actual power (P) is the product of apparent power (S) and the power factor:

$$P = E \cdot I \cdot p_f$$

c) Power factor:

$$p_f = \frac{P}{S}$$

where

p_f = power factor
P = real power (W)
S = apparent power (VA)

If φ is the phase angle between the current and the voltage, then the power factor is equal to $\left| \cos \phi \right|$ and the real power is

$$P = S \cos \phi$$

d) Reactive power
Reactive power (Q) is the power returned to the source by the reactive components of the circuit:

$$Q = I_L^2 X_l - I_C^2 X_C$$

where

Q = reactive power (VAr)

I_L = inductive current (A)

I_C = capacitive current (A)

X_L = inductive reactance (Ω)

X_C = capacitive reactance (Ω)

43. Parallel Circuit

There is one major difference between a series circuit and a parallel circuit. The difference is that the current is the same in all parts of a series circuit, whereas voltage is the same across all branches of a parallel circuit. Because of this difference, the total impedance of a parallel circuit must be computed on the basis of the current in the circuit.

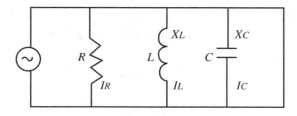

When working with a parallel circuit, one must use the following formulas:

a) Voltage

Voltage is the same across all branches of a parallel circuit. Thus,

$$E = E_R = E_L = E_C$$

---

where

E = total voltage across circuit (V)

E_L = inductive voltage (V)

E_R = resistance voltage (V)

E_C = capacitive voltage (V)

b) Current:

$$I_Z = \sqrt{I_R^2 + I_X^2} = \sqrt{I_R^2 + \left(I_L - I_C\right)^2}$$

$$I_X = I_L - I_C$$

where

I_Z = impedance current (A)

I_R = resistance current (A)

I_L = capacitive current (A)

c) Impedance

The impedance Z of a parallel circuit is found by the formula

$$Z = \frac{E}{I_Z} = \frac{E}{\sqrt{I_R^2 + I_X^2}}$$

LIGHT

In a strict sense, light is the region of the electromagnetic spectrum that can be perceived by human vision. That is, it is the visible spectrum, which includes wavelengths ranging approximately from 0.4 micrometers to 0.7 micrometers.

This section contains the most frequently used formulas, rules and definitions relating to the following:

- General Terms
- Photometry
- Reflection, Refraction, and Polarization
- Geometric Optics

LIGHT
General Terms

---

1. Visible Light
Visible light is the portion of the electromagnetic spectrum between the frequencies of 4.3×10^{14} Hz and 7.5×10^{14} Hz. Hence,

$$4.3 \times 10^{14} \leq f \leq 7.5 \times 10^{14} \text{ (Hz)}$$

2. Speed of Light
The speed of light is a scalar quantity, having only magnitude but no direction. The following basic relationship exists for all electromagnetic waves, and relates the frequency, wavelength, and speed of the waves:

$$c = \lambda f$$

where

c = speed of light, 3.00×10^{8} (m/s)
f = frequency (Hz)
λ = wavelength (m).

3. Light as a Particle
In quantum theory, particles of light are given the name "photons." A photon has energy defined by the equation

$$E = hf = \frac{hc}{\lambda}$$

LIGHT
Photometry

---

where

E = energy (J)

h = Planck's constant, $h = 6.62 \times 10^{-34}$ (J·s)

f = frequency (Hz)

λ = wavelength (m)

c = speed of light, 3.00×10^{8} (m/s).

4. Luminous Intensity

Luminous intensity refers to the amount of luminous flux emitted into a solid angle of space in a specified direction:

$$I_v = \frac{r^2 E_v}{\cos \theta}$$

where

I_v = luminous intensity (cd)

r = distance between the source and the surface (m)

E_v = illuminance (lux).

5. Luminous Flux

Luminous flux is a measure of the energy emitted by a light source in all directions:

$$\Phi_v = \Omega I_v$$

where

Φ_v = luminous flux (lm)

Ω = solid angle (sr)

I_v = luminous intensity (cd).

6. Luminous Energy

Luminous energy is photometrically weighted radiant energy:

$$Q_v = \Phi_v t$$

where

Q_v = luminous energy (lm·s)

Φ_v = luminous flux (lm)

t = time (s).

7. Illuminance

Illuminance is the luminous flux collected by a unit of a surface:

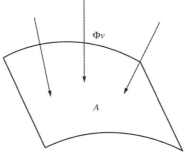

$$E_v = \frac{\Phi_v}{A} = \frac{\Omega I_v}{A}$$

where

E_v = illuminance (lx)

LIGHT
Photometry

Φ_v = luminous flux (lm)

Ω = solid angle (sr)

I_v = luminous intensity (cd)

A = surface area(m^2).

8. Luminance

Luminance is the luminous intensity emitted by the surface area of one square meter of the light source. The luminance value indicates the amount of glare and discomfort when we look at a lighting source. The following figure shows the concept:

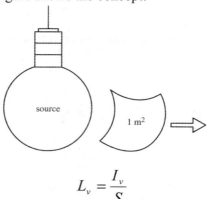

$$L_v = \frac{I_v}{S}$$

where

L_v = luminance $\left(cd/m^2\right)$

I_v = luminous intensity (cd)

S = area of the source surface perpendicular to the given direction $\left(m^2\right)$.

LIGHT
Reflection, Refraction, and Polarization

---

9. Laws of Reflection

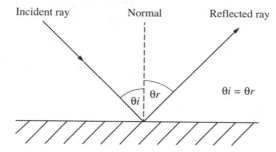

A ray of light is a line whose direction gives the direction of flow of radiant energy.

a) First law of reflection
The angle of incidence is equal to the angle of reflection. That is,

$$\theta_i = \theta_r$$

where

θ_i = angle of incidence

θ_r = angle of reflection

b) Second law of reflection
The incident ray, the reflected ray, and the normal to the surface all lie in the same plane.

10. Refraction
In an isotropic medium, rays are strength lines, along which energy travels at speed:

$$v = \frac{c}{n}$$

LIGHT
Reflection, Refraction, and Polarization

---

where

    n = refractive index of the medium
    c = speed of light in vacuum (m/s)

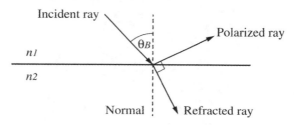

a) Law of refraction

When a ray of light passes at an angle from a medium of lower optical density to a denser medium, the light ray is bent toward the normal.

When a ray of a light passes, at an angle, from a denser medium to one less dense, the light is bent away from the normal. Hence,

$$\frac{\sin \theta_i}{\sin \theta_2} = \frac{n_2}{n_1}$$

$$n_1 = \frac{c}{v_1}, \qquad n_2 = \frac{c}{v_2}, \qquad \frac{n_2}{n_1} = \frac{v_1}{v_2},$$

where

    v_1 = speed of light in medium 1 (m/s)
    v_2 = speed of light in medium 2 (m/s)
    n_1 = refractive index of medium 1

LIGHT
Reflection, Refraction, and Polarization

n_2 = refractive index of medium 2

c = speed of light in vacuum (m/s)

If $n_1 > n_2$ and θ_i exceeds the critical θ_c, where

$$\sin \theta_c = \frac{n_2}{n_1},$$

then there will be no refracted ray; this is a phenomenon called *total reflection*.

11. Polarization

An electromagnetic or other transverse wave is polarized whenever the disturbance lacks cylindrical symmetry about the ray direction.

When the reflected ray is at 90° to the refracted ray, the transverse component of the electric field lies along the path of the reflection.

This would make the wave longitudinal, so clearly there is no transverse component in the reflection.

The incident angle at which this happens is called the polarizing angle or Brewster's angle:

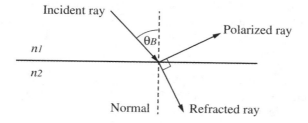

LIGHT
Geometrical Optics

$$\tan \theta_B = \frac{n_2}{n_1}$$

where

θ_B = Brewster's angle (°)

n_1 = refractive index of the incident medium

n_2 = refractive index of the reflecting medium.

12. Plane Mirrors

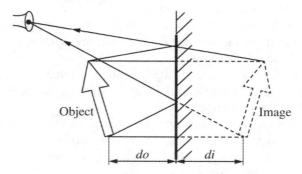

The image is at the same distance behind the mirror as the object is in front of it:

$$d_o = d_i$$

13. Concave Mirrors

Depending upon the position of the object, the image will be real or virtual.

LIGHT
Geometrical Optics

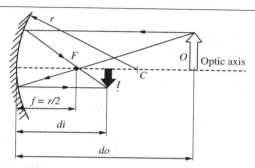

14. Convex Mirrors

Convex mirrors produce only virtual and smaller images.

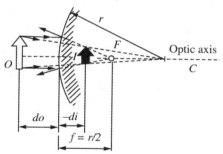

15. Mirror Formula

$$\frac{1}{f} = \frac{1}{d_o} + \frac{1}{d_i}; \qquad \frac{h_1}{h_o} = \frac{d_i}{d}$$

where

f = focal length of mirror
d_o = distance of object from mirror
d_i = distance of image from mirror

LIGHT
Geometrical Optics

---

h_i = image height
h_o = object height

16. Lens Equation

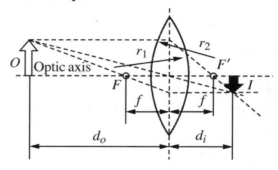

$$\frac{1}{f} = \frac{1}{d_i} + \frac{1}{d_o} = (n-1)\left(\frac{1}{r_1} + \frac{1}{r_2}\right); \quad m = \frac{h_i}{h_o} = \frac{d_i}{d_o}$$

where

f = focal length
F, F' = focuses
r_1, r_2 = radii of curvatures
n = refractive index
h_i = image height
h_o = object height
m = magnification factor
d_o = object distance from lens center
d_i = image distance from lens center.

WAVE MOTION AND SOUND

Wave motion is defined as the movement of a distortion of a material or medium, where the individual parts or elements of the material only move back and forth, up and down, or in a cyclical pattern.

This section contains the most frequently used formulas, rules, and definitions relating to the following:

- Wave Terminology
- Wave Phenomena
- Electromagnetic Waves, Energy, and Spectrum
- Sound Waves and Speed

WAVE MOTION AND SOUND
Wave Terminology

1. Definitions and Graphs

A wave is a transfer of energy, in the form of a disturbance, through some medium, but without translocation of the medium.

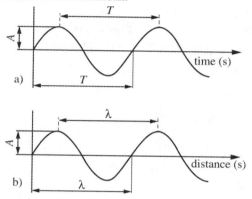

Waves may be graphed as a function of time (a) or function of distance (b). A single-frequency wave will appear as a sine wave in either case. From the distance graph, the wavelength may be determined. From the time graph, the period and frequency can be obtained. From both together, the wave speed can be determined.

2. Wavelength

Wavelength λ is defined as the distance from one crest (or maximum) of the wave to the next crest.

WAVE MOTION AND SOUND
Wave Terminology

---

3. Amplitude

The amplitude A of a wave is the maximum displacement from the equilibrium or rest position.

4. Velocity

The velocity v of the wave is the measurement of how fast a crest is moving from a fixed point:

$$v = \frac{\lambda}{T} = \lambda f$$

where

v = velocity (m/s)
T = period (s)
f = frequency (1/s or Hz)
λ = wavelength (m).

5. Frequency

The frequency f of waves is the rate at which the crests or peaks pass a given point:

$$f = \frac{1}{T}$$

6. Period

The period T is the time required to complete one full cycle:

$$T = \frac{1}{f}$$

WAVE MOTION AND SOUND
Wave Phenomena

---

7. Wave on a Stretched String

The speed of a wave traveling on a stretched uniform string is given by

$$v = \sqrt{\frac{F}{\rho}}$$

where

F = tension in the string
ρ = linear density of the string.

8. The Sinusoidal Wave

The sinusoidal wave is a periodic wave described by a function of two variables:

$$y(x,t) = A\cos\left[k(x - vt)\right]$$

where

A = amplitude
k = angular wave number
v = wave speed

a) Wave speed:

$$v = \frac{\omega}{k}$$

b) Period

For a particular x, y is a periodic function of t with period:

WAVE MOTION AND SOUND
Wave Phenomena

$$T = \frac{2\pi}{\omega}$$

c) Wavelength

For a particular t, function y is a periodic function of x, with the wavelength given by

$$\lambda = \frac{2\pi}{k}$$

d) Power

The average power transmitted by a sinusoidal wave can be calculated by the formula

$$P_{avg} = \frac{1}{2}\omega^2 A^2 \rho v$$

where

    A = amplitude

    ρ = density of a medium

    v = wave speed.

    ω = angular frequency

e) Energy

For a wave on string, the energy per unit length is given by

$$E_l = \frac{P_{avg}}{v}$$

where

    P_{avg} = average power transmitted by the wave

    v = wave speed.

WAVE MOTION AND SOUND
Electromagnetic Waves, Energy, and Spectrum

---

9. Electromagnetic Waves

These waves are made up of electric and magnetic fields whose strengths oscillate at the same frequency and phase. Unlike mechanical waves, which require a medium in order to transport their energy, electromagnetic waves are capable of traveling through a vacuum.

Although they seem different, radio waves, microwaves, X-rays, and even visible light are all waves of energy called electromagnetic waves.

Electromagnetic waves have amplitude, wavelength, velocity, and frequency. The creation and detection of the wave depend on the range of wavelengths.

a) Wave speed:

$$v = c = \lambda f = \frac{\lambda}{T}$$

where

c = speed of light (3.00×10^8 m/s)
f = frequency (1/s)
λ = wavelength (m)
T = period (s).

10. Electromagnetic Energy

Electromagnetic energy at a particular wavelength λ (in vacuum) has an associated frequency f and photon energy E:

$$E = h \cdot f$$

WAVE MOTION AND SOUND
Electromagnetic Waves, Energy, and Spectrum

where

h = Planck's constant = 6.62607×10^{-34} (J·s)
f = frequency (1/s).

11. The Electromagnetic Spectrum

The electromagnetic spectrum is a continuum of all electromagnetic waves arranged according to frequency and wavelength, as shown below.

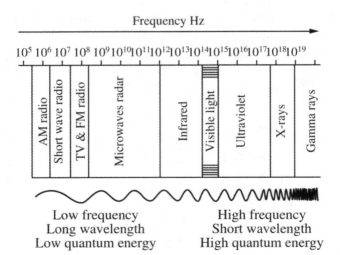

WAVE MOTION AND SOUND
Sound Waves and Speed

12. Sound Waves

Sound is a longitudinal wave in a medium created by the vibration of some object:

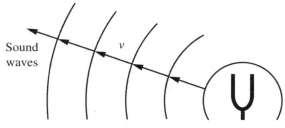

Tuning fork

13. Speed of Sound in Air

The speed in dry air at 1 atmosphere pressure and $0°$ C is 331.4 m/s. Changes in humidity and temperature cause a variation in the speed of sound. The speed of sound increases with temperature at the rate of 0.61 m/s $°$C. The speed of sound in dry air at 1 atmosphere pressure is then given by

$$v = 331.4 + (0.610) \cdot t_C$$

where

t_C = air temperature ($°$C).

14. Sound Speed in Gases

The speed of sound in an ideal gas is given by the formula

$$v = \sqrt{\frac{\gamma R T}{M}}$$

where

v = speed of sound (m/s)

R = universal gas constant = 8.314 J/mol

$\text{K}T$ = absolute temperature (K)

M = molecular mass of gas (kg/mol)

γ = adiabatic constant

For air, the adiabatic constant is $\gamma = 1.4$ and the average molecular mass (M) for dry air is 28.95 g/mol. Hence,

$$v = \sqrt{\frac{\gamma R T}{M}} = \sqrt{\frac{1.4(8.314)T}{0.02895}} = 20.05\sqrt{T} \ \ (\text{m/s})$$

15. The Doppler Effect

Suppose that a source emitting sound waves of frequency f_s and an observer move along the same straight line. Then the observer will hear sound of the frequency

$$f_o = f_s \frac{v \pm v_o}{v \mp v_s}$$

where

f_s = the source sound frequency
f_o = the observer sound frequency
v_o = the relative speed of the observer
v_s = the relative speed of the source
v = the sound speed in this medium

The choice of using a plus (+) or minus (−) sign is made according to the convention that if the source and observer are moving towards each other the observer frequency f_o is higher than the actual frequency f_s. Likewise, if the source and observer are moving away from each other, the observer frequency f_o is lower than the actual frequency f_s.

APPENDIX

Fundamental Physical Constants

Name	Symbol and Value
alpha particle mass	$m_\alpha = 6.6446565 \times 10^{-27}$ kg
atomic mass constant	$m_u = 1.66053886 \times 10^{-27}$ kg
atomic unit of energy	$E_h = 4.35974417 \times 10^{-18}$ J
atomic unit of force	$E_h / a_0 = 8.2387225 \times 10^{-8}$ N
atomic unit of length	$a_0 = 0.5291772108 \times 10^{-10}$ m
atomic unit of mass	$m_e = 9.1093826 \times 10^{-31}$ kg
Avogadro's constant	$N_A = 6.0221415 \times 10^{23}$ mol^{-1}
Bohr radius	$a_0 = 0.5291772108 \times 10^{-10}$ m
Boltzmann constant	$k_B = 1.3806505 \times 10^{-23}$ J K^{-1}
classical electron radius	$r_e = 2.817940325 \times 10^{-15}$ m
elementary charge	$e = 1.60217653 \times 10^{-19}$ C
electron charge to mass	$\dfrac{e}{m_e} = 1.758\ 82012 \times 10^{11}$ C/kg

Continued

electron rest mass	$m_e = 9.11 \times 10^{-31}$ kg or 0.511 MeV
electron gyromagnetic ratio	$\gamma_e = 1.76085974 \times 10^{11}$ s^{-1} T^{-1}
electron magnetic moment	$\mu_e = -928.476412 \times 10^{-26}$ J T^{-1}
electron g factor	$g_e = -2.0023193043718$
Faraday's constant	$F = 96485.3383$ C mol^{-1}
fine-structure constant	$\alpha = 7.297352568 \times 10^{-3}$
gas constant	$R = 8.31 \times 10^3$ J K^{-1}kmol^1
gas (ideal) normal volume	$V_o = 22.4$ m^{3}kmol^{-1}
Hydrogen atom (rest mass)	$m_H = 1.673 \times 10^{-27}$ kg or 938.8 MeV
molar mass constant	$M_u = 1 \times 10^{-3}$ kg mol^{-1}
molar volume of ideal gas	$V_m = 22.710981 \times 10^{-3}$ m^3 mol^{-1}
neutron g factor	$g_n = -3.82608546$
neutron gyromagnetic ratio	$\gamma_n = 1.83247183 \times 10^8$ s^{-1} T^{-1}
neutron mass	$m_n = 1.67492728 \times 10^{-27}$ kg

Continued

Newtonian constant of gravitation	$G = 6.6742 \times 10^{-11} \text{ m}^3 \text{ kg}^{-1} \text{ s}^{-2}$
Nuclear magneton	$\mu_N = 5.05078343 \times 10^{-27} \text{ J T}^{-1}$
Planck's constant	$h = 6.6260693 \times 10^{-34} \text{ J s}$
Planck mass	$m_P = 2.17645 \times 10^{-8} \text{ kg}$
proton charge to mass quotient	$\dfrac{e}{m_p} = 9.57883376 \times 10^{7} \text{ C kg}^{-1}$
proton g factor	$g_p = 5.585694701$
proton gyromagnetic ratio	$\gamma_p = 2.67522205 \times 10^{8} \text{ s}^{-1} \text{ T}^{-1}$
proton mass	$m_p = 1.67262171 \times 10^{-27} \text{ kg}$
proton-electron mass ratio	$\dfrac{m_p}{m_e} = 1836.15267261$
speed of light in vacuum	$c = 299792458 \text{ m s}^{-1}$
standard acceleration of gravity	$g = 9.80665 \text{ m s}^{-2}$
standard atmosphere	$p = 101325 \text{ Pa}$
Stefan-Boltzmann constant	$\sigma = 5.670400 \times 10^{-8} \text{ W m}^{-2} \text{ K}^{-4}$

INDEX

433

ABOUT THE AUTHOR

Vukota Boljanovic, Ph.D., has more than 35 years of experience in applied engineering in the aircraft and automotive industries as well as in education. He received his B.S., M.S., and Ph.D. in Mechanical Engineering and worked in Europe for many years in both academia and industry, including as a Professor of Aerospace Engineering and later as Vice President for Research and Development for a major aircraft company. He also spent more than five years in the American automotive industry. He has performed extensive research in the development of advanced technology and materials and has designed tools and dies for the aircraft and automotive industries.

Boljanovic also is the author of multiple books, including seven books with Industrial Press, Inc., a number of which have been translated into other languages. He also has contributed updated practical material on punches, dies, and presswork; sheet-metal working and presses; and powder metallurgy to three editions of the publisher's flagship product, *Machinery's Handbook*. Boljanovic is widely recognized by both academia and industry for his contributions to manufacturing, engineering, and applied mathematics.